Economic Games and Strategic Behaviour

For Astrid Annette

Economic Games and Strategic Behaviour

Theory and Application

Frank Stähler

*Senior Lecturer, Institute of Economic Theory,
University of Kiel, Germany*

Edward Elgar
Cheltenham, UK • Northampton, MA, USA

Published by
Edward Elgar Publishing Limited
8 Lansdown Place
Cheltenham
Glos GL50 2HU
UK

Edward Elgar Publishing, Inc.
6 Market Street
Northampton
Massachusetts 01060
USA

A catalogue record for this book
is available from the British Library

Library of Congress Cataloguing in Publication Data

Stähler, Frank.
    Economic games and strategic behaviour: theory and application /
Frank Stähler.
    Includes bibliographical references and index.
    1. Game theory. I. Title.
HB144.S73   1998
    330'.01'5193—dc21   98–13458
                                                                         CIP

ISBN 1 85898 905 1

Printed and bound in Great Britain by Biddles Ltd, Guildford and King's Lynn

Contents

Figures

Tables

Preface

This book deals with economic games and strategic behaviour. It will introduce a model which can determine a solution for economic games which are subject to repeated play. My interest for long-term cooperation was shaped during a research project for the Volkswagen Foundation which dealt with self-enforcing environmental agreements. The problem of finding an agreement which is unanimously accepted, led me to strategic bargaining theory, the need for immunity against defection led me to the theory of repeated games. Since both theories employ the discount factor as the explaining variable, I decided to attempt to combine both approaches; this book is the result.

The success of such an attempt depends on the quality of ideas presented in the book, on which readers can make their own judgements and for which the author bears full responsibility. However, a necessary condition for a book to be successful is that it is written, and that a book is written depends not only on the author but also substantially on the environment in which it is written. This work was done while I was with the Kiel Institute of World Economics. It has been accepted as a *Habilitationsschrift* by the Department of Economics of the University of Kiel. Parts of the book were presented in seminars at the University of Kiel and the University of Frankfurt (Oder). I am indebted to many people for their comments, criticisms and suggestions. Among them are Friedel Bolle, Wolfgang Buchholz, Frank Bulthaupt, Johannes Heister, Gernot Klepper, Ernst Mohr and Jonathan Thomas.

I am especially grateful to Horst Herberg, Horst Siebert and Hans-Werner Wohltmann, who read with great care a first draft of the manuscript, in which my ideas were rather obscure, and who got never tired of discussing changes, improvements and potential applications. Horst Herberg advised me to make my arguments more understandable. He also found out that one of the compliance constraints which my model employs is redundant. Horst Siebert encouraged me to discuss applications in detail. Hans-Werner Wohltmann made very useful suggestions for improving my mathematical exposition and for making the dynamic assumptions of my model more transparent. Without all these comments and suggestions, the book would suffer from several deficiencies. Of course, any remaining errors are in my responsibility.

Last but not least I wish to thank my family. My sons David and Leander demonstrated impressively and refreshingly that there are other games to explore and – more importantly – to play which provided a lot of fun right from the beginning (and they still do). Their father enjoyed it very much (and still does). Words cannot express adequately the many thanks I owe to my wife. She tolerated with great patience my fluctuating moods, which were strongly correlated to the importance of the work as it was perceived by me. This book is dedicated to her.

1. Preview

1.1 THE PROBLEM

In 1947, the discussion on modelling economic interactions was substantially stimulated by the publication of John von Neumann and Oscar Morgenstern's book entitled *Theory of Games and Economic Behavior*. This book was the starting-point for considering economics from a strictly game-theoretic perspective. Nowadays, the examination of economic games is at the heart of contemporary economics, and certain theorems and concepts, for example those by Nash and Selten, have become textbook economics. This book deals with a certain part of this field and explores the relationship between a certain class of economic games and the strategic behaviour of players. It is the understanding of this book that the important relevant economic games are those in which uncoordinated behaviour does not lead to efficient results. Hence, the focus of this book is on economic games in which a certain strategic behaviour may be able to overcome potential inefficiencies through coordination among players. The coordination device of an economic game will be specified by a contract.

Contracts serve for strategy specification. Since its early beginnings, economics has stressed the importance of coordinated behaviour for the welfare of an economy. When the actions of all parties can be coordinated, every party is expected to be better off by coordinated rather than by uncoordinated behaviour. This coordination may either be provided by institutional arrangements such as the coordination in the marketplace of a market economy or is arrived at by an agreement among all parties involved. Such an agreement may be in the explicit form of a contract. A contract specifies declarations of intent of all parties, which do not contradict each other. However, contracts are only reliable if all parties expect that every party will comply with the terms of the contract. Otherwise, coordination is at risk and realizing mutual benefits from coordinated behaviour is not guaranteed. These risks are a real threat for a contract if a unilateral deviation from the terms of the contract benefits the deviating party more than if they had fulfilled the contract. This case applies if the problem under consideration is of the prisoners' dilemma type.

Contracts are reliable at least to a certain extent if they are enforced by a third party or if a third party effectively and credibly punishes any deviation of a party from the terms of the contract. In many cases, however, no third party exists or is too costly to be made use of. One may distinguish two sets of contracts which cannot rely on external enforcement. International contracts are always subject to compliance risks because the sovereignty status of every country may prevent countries from committing themselves credibly to a certain policy. Private contracts are subject to compliance risks if the enforcement of third parties is not guaranteed or enforcement is too costly. When external enforcement is not possible or is too costly, these contracts must be self-enforcing. A self-enforcing contract is a contract which does not rely on external threats but on threats which are laid down in the contract itself. The threats laid down in the contract must be credible and must ensure that a party complies with the terms of the contract.

The actions which respond to a contract breach typically affect the future benefits of all parties. Hence, a self-enforcing contract must rely on an intertemporal estimate: every party computes the current benefits of not keeping the terms of the contract and the costs which may arise from future responses to this behaviour. When the threat is credible and severe, one may expect that every party will refrain from breaking the contracts. The severity of future threats depends crucially on the weight a party gives to future benefits. The stronger the influence of future actions on current behaviour, the stronger the potential future threat which responds to deviant behaviour.

There are many examples for contracts which must be self-enforcing to be reliable. In trade theory, at least two examples are well known which mirror coordination problems of the prisoners' dilemma type. First, a large country may improve unilaterally on the free trade equilibrium by imposing an import tariff which improves its terms of trade. When both countries impose a tariff, they are both likely to be worse off compared to the free trade equilibrium. Second, strategic trade theory has demonstrated that a country may improve the profits of a domestic industry by unilaterally subsidizing production or research and development. When both countries provide subsidies, they are worse off than when there are no subsidies. In both cases, the sovereignty status of both countries may prevent their credible commitment not to introduce the respective instruments. Hence, any trade policy agreement must be self-enforcing so that the unilateral introduction of trade policy instruments which conflicts with the terms of the contract must be unprofitable for each country because the current welfare improvements do not exceed the welfare losses which are due to the future responses to the contract breach.

Another example is the management of international environmental problems, which are often due to transboundary pollution from one country to another. In these cases, both countries decrease their environmental damages by jointly reducing national pollution. However, compared to joint environmental policies, each country is better off if only the other country reduces pollution. Since neither country can credibly commit itself not to deviate from a certain policy, an international contract must guarantee that unilateral defection from the agreed-upon pollution reduction levels does not pay because the current gains should not exceed the costs which result from future responses. In fact, all international contracts which are subject to sovereignty constraints must be self-enforcing in this sense. The problems which arise when this constraint is violated were impressively demonstrated by the debt crisis. During the debt crisis, sovereign countries refused to pay debt services to foreign banks. Obviously, certain debt contracts had not taken into account the possibility that sovereign countries may repudiate debt services.

Self-enforcing contracts are not restricted to international coordination problems. They may also be used when either coordination cannot be laid down by an enforceable contract or enforcement is too costly. The first case applies when coordination among suppliers in an industrial market may lead to a certain degree of collusion. Since collusion is supposed to lower consumer rents to a higher extent than it increases producer rents, antitrust laws do not in general allow market behaviour of firms to be coordinated by an explicit agreement. In the standard cases, collusion benefits all firms compared to non-cooperation but each firm is even better off when all other firms collude but it does not. Hence, all firms are in a prisoners' dilemma situation because any explicit collusive agreement conflicts with antitrust laws. Then, coordination may take place only through a self-enforcing contract which specifies future actions such that every firm will refrain from breaking this contract.

The second case refers to prohibitively high enforcement costs. Consider for example a buyer and a seller. Both may agree upon a certain contract which specifies the terms of the delivery and the terms of the payment. The buyer may break the contract by not paying the seller, and the seller may break the contract by not delivering. Contract breaches are a real threat if bringing a breach of contract to court is too costly. Costs may arise because enforcement or compensation will be delayed, or because each party has to bear some administrative costs. If these costs are too high compared to the potential gains from the court's decision, a contract will be designed as a self-enforcing contract. Then, the contract specifies what actions will be taken in the future in the case of defection.

In every case, self-enforcing contracts rely on the impact of current behaviour on long-run cooperation. This book explores the role of self-enforcing contracts in a stationary world: the models assume that neither the action spaces of agents nor the payoff functions change in the course of time. Thereby, the analysis is restricted to situations which are repeated or are expected to be repeated with a certain probability. Stationarity of the environment is of course a restrictive assumption because it neglects feedbacks of current actions on future options. Hence, one should keep in mind that the model and the results do not apply to all conceivable coordination problems. All problems which trigger general equilibrium effects so that stock variables are permanently changed cannot be modelled as repeated situations but imply dynamic problems. We shall turn to this issue in the last chapter when discussing an agenda for future research.

This book discusses strategic bargaining for self-enforcing contracts. It will show that strategic bargaining determines the self-enforcing contract which is deliberately chosen by two agents and which specifies a solution for a certain class of economic games. In general, decision makers will be referred to as agents, irrespective of whether they are individuals or representatives of an organization. Before deriving the strategic bargaining solution, the notions of a self-enforcing contract and of strategic bargaining must be clarified. Then, a model has to be developed which explains strategic bargaining in this context. Finally, several features of the model will be discussed. In particular, the following questions will be answered:

- What role do contracts play in a non-cooperative environment in which the economic game does not guarantee compliance?

    In a non-cooperative environment, no third party is able or willing to enforce a contract. A contract will be referred to as two consistent declarations of intent of two agents. Then, contracts are not restricted to those contracts which are enforced by third parties when a party intends to break the contract unilaterally. Hence, we shall investigate what contracts look like in a non-cooperative environment.
- What conditions should a contract fulfil to qualify as self-enforcing, that is, for an instrument which sustains long-run cooperation?

    When contracts serve for specifying actions to be taken by the parties, self-enforcing contracts must also specify what must be done in the case of non-compliance of a party because recourse to external enforcement cannot be taken. The instruments which respond to non-compliance of a party should be credible, and a certain degree of credibility implies certain conditions which a self-enforcing contract should fulfil.
- Which self-enforcing contract will be chosen?

    When several self-enforcing contracts exist which cannot be Pareto ranked, a selection problem arises. A strategic bargaining approach will cope with this problem and will demonstrate which contract will be chosen. The solution will be unique.

- How does the institutional setting influence the outcome?

  When the institutional setting decides the outcome, it is interesting to explore whether an agent improves the outcome for his/her partner or him-/herself when he/she changes the institutional setting.

- Is a Pareto-optimal contract always chosen if possible?

  When the threat of future actions which respond to deviating behaviour is severe, one may expect that outcomes may be credibly agreed upon which would also be agreed upon if contracts were enforceable because a Pareto-optimal contract maximizes total payoffs. This case deserves special attention because the chosen contract will be shown to result from strategic bargaining and not just from feasibility.

- Do transfers always improve on the corresponding outcome without transfers?

  Transfers between agents are often supposed to support cooperation in a non-cooperative environment because they are thought to reconcile efficiency and compliance. Their role will be discussed under the consistent assumption that they are also subject to compliance risks.

- Might a party prefer a self-enforcing contract even if enforceable contracts were possible?

  Non-compliance of a partner can only be avoided if the agreement gives him/her sufficiently high cooperation gains compared to non-compliance. Then, it cannot be ruled out a priori that this agent is better off with a self-enforcing contract which guarantees him/her a certain number of gains which may be subject to bargaining for an enforceable contract.

All these questions will be answered in this book, and will be summarized in the concluding remarks of Chapter 6.

1.2 OBJECTIVE AND THEORETICAL FOUNDATIONS

The scope for long-run cooperation in a non-cooperative environment depends on the strategies of agents and the strategic interactions. During the last decades, economics has explained strategic interactions between a small number of individuals by using the tools of game theory. Game theory has enabled economists to explore the impacts of strategic interactions on economic performances. This book deals with some of these developments. It will discuss and employ the theory of repeated games and strategic bargaining theory. The theory of repeated games has demonstrated that a scheme other than the one-shot equilibria of stage games may also give long-run equilibria. This theory demonstrated that Pareto-inferior non-cooperative outcomes may be overcome if repetition plays a disciplinary role for cooperation. In this setting, any agreement must be self-enforcing because no third party is able to enforce any coordinated behaviour that both agents have agreed upon. This theory shows that future cooperation is given more weight for current action the more patient the agents involved are. Strategic bargaining theory determines the agreed-upon outcome in a purely non-cooperative setting. In that setting, enforceability of an agreement is at no risk but no third party is able to enforce a certain division of bargaining gains. Hence, strategic bargaining theory deals with the problem of finding and not of enforcing an agreement. It has explained the division of bargaining gains in a bilateral setting when a fairly general alternating offer structure is assumed. In that setting, a relatively impatient player is given less bargaining power because he/she is willing to sacrifice substantial bargaining gains in order to reach an early agreement.

Both theories employ and depend crucially on the discount factors of agents: the lower the discount factor, the less weight is given to future responses to current actions, and the less is the bargaining power. Repeated game theory proves that a scheme other than one-shot equilibria may be sustainable but cannot disprove the relevance of one-shot equilibria. The set of attainable equilibria is enlarged for the price of an ambivalent result. The strategic bargaining approach determines a unique solution for dividing bargaining gains for only one possible realization of bargaining gains. To my knowledge, there has been no attempt to link both theories although they both employ the discount factor as an explanatory variable. Only the papers by Okada (1991a,b) deal with long-term aspects of bargaining in a non-cooperative environment by comparing short-term and long-term enforceability. This book will present an approach which combines both theories in a completely non-cooperative environment, and will thereby increase the explanatory power of both theories:

- Strategic bargaining theory assumes in standard bargaining models that no agreement which results from the bargaining process is at risk because every unanimously agreed-upon division of bargaining gains is enforceable. This assumption will be relinquished. Instead, any agreement reached by mutual bargaining is assumed to be self-enforcing so that future responses to not agreed-upon behaviour must prevent a current breach of the agreement.
- Repeated game theory and the Folk Theorem proved that every attainable equilibrium, that is, the non-cooperative outcome, the

cooperative outcome and everything 'in between', may give an equilibrium. The ambiguity problem will be solved. These equilibria, if attainable, give the concerned agents different utilities. Different utilities for different agreements will be addressed by a well-defined bargaining procedure which determines a unique bargaining-based solution. The set of equilibria which are candidates for the bargaining-based solution will be referred to as the set of self-enforcing contracts.

- Both theories often assume identical discount factors for all agents. This assumption will be abandoned. Instead, the impacts of different discount factors will be explored.

The aim of this book is, however, not purely theoretical in that it resolves the ambiguity problem of repeated games and adds an application to strategic bargaining theory. It is not only addressed to specialists in repeated games and strategic bargaining theory but is also written for economists who are interested in or familiar with strategic bargaining theory, contract theory, the theory of repeated games and applied economics. Addressing such a heterogeneous audience requires that the formal structure of modelling is comprehensible to all potential readers. The compromises made by the models are overviewed in Section 1.4.

1.3 ORGANIZATION OF THE BOOK

This chapter is followed by five chapters:

- Chapter 2 will link up the theory of repeated games and strategic bargaining theory. It will develop the essentials of both theories which give the ingredients for a theory of strategic bargaining for self-enforcing contracts. This theory will be able to explain simultaneously the division and the enforceability of bargaining gains in a non-cooperative environment.
- Chapters 3, 4 and 5 will discuss different applications of strategic bargaining for self-enforcing contracts. Chapters 3 and 4 will both deal with bargaining for a collective good. Chapter 3 will ignore any transfers between agents, Chapter 4 will allow for transfers. Chapter 5 will deal with a typical buyer–seller case. The stage game models in all chapters will be of the prisoners' dilemma type. Furthermore, every chapter will discuss compliance constraints and optimal policies for every agent, will develop a so-called concession line and will determine the strategic bargaining solution. Since the impact of discount factors is twofold,

variations of discount factors will be addressed by computing marginal changes of bargaining solutions. The impacts on selected applications will be given in Chapters 3 and 4. A summary of results will be presented at the end of every chapter.

- Chapter 6 will conclude this book, will discuss an agenda for future research, and will give some policy implications which will incorporate some of the arguments developed in different chapters in order to shed some light on policy design. Discussion of policy implications is usually not on the agenda when strategic bargaining models are. However, the assumptions which distinguish Chapters 3, 4, 5 and bargaining for enforceable contracts are basically assumptions concerning the institutional setting of strategic bargaining. Reconsidering some selected results of these chapters from the viewpoint of designing institutional settings will give some straightforward policy implications.

The book will demonstrate that strategic bargaining for self-enforcing contracts solves the indeterminacy problem of repeated games on the one hand and enriches strategic bargaining theory on the other. Because of the twofold impact of discount factors, their role for distributing bargaining gains will be shown to be ambiguous in general. Thus, a relatively high level of patience does not necessarily benefit an agent. In other words, the preference of an agent to bargain with different types of agents cannot be ranked along with the other agent's discount factor. This result may also help to explain team selection.

1.4 SOME METHODOLOGICAL REMARKS

This book develops the link between strategic bargaining theory and the theory of self-enforcing contracts, two theories which have so far evolved separately. Linking up both theories should give some insight into the design of long-term relationships between two agents. In order to develop the results straightforwardly, however, some restricting assumptions are necessary. The first assumption restricts the analysis to two agents. Since its early beginnings, strategic bargaining theory has concentrated mainly on relationships between two agents. Enlarging the number of bargaining agents increases the complexity enormously and makes results more crucially dependent on the bargaining rules (see, for example, Binmore, 1985). Krishna and Serrano (1996) have shown that employing a certain axiom makes multilateral bargaining explainable using a basically strategic bargaining model.

The discussion of repeated games which will serve for defining self-enforcing contracts took a similar path because early models ignored the potential difference between two- and *n*-agent models by assuming through appropriate strategy specification that the results of the basic model carry over to *n*-agent models. Since the notions of coalition formation and coalition-proofness have entered the discussion, the potential role of different group sizes has been acknowledged (for a first paper, see Bernheim, Peleg and Whinston, 1987). The results of bilateral relationships do not carry over to multilateral relationships if coalitions are allowed. Consider, for example, three agents operating in a non-cooperative environment. Two out of these three agents may form a coalition which may coordinate their joint actions without including the third agent. Obviously, all options of all possible coalitions must be taken into consideration when discussing coordinated behaviour among all agents. This requirement increases the complexity enormously and implies ambiguous results unless a certain division of bargaining gains is assumed (see Stähler, 1996a). The standard case of only two agents avoids this ambiguity.

Strategic bargaining for a self-enforcing contract between two agents has already turned out to be a very complex case which deserves its own approach. Additionally, the case of two agents is a very important one. In many cases, interactions between oligopolists or between countries are basically conflicts between two parties. In the case of several parties, discussing the case of two agents is a prerequisite in order to be able to determine the role of coalitions. Thus, the two-agent case must be dealt with explicitly and serves at least as a potential starting-point for further research.

The second assumption is the already mentioned stationarity assumption which repeated games employ. Repeated games are a very specific variant of dynamic games because they assume that the payoff structure remains unchanged in the course of time. Alternatively, a more realistic assumption would specify that the conditions which determine the payoffs of agents change in the course of time due to changes in the environment and changes which are the results of deliberate behaviour of agents, for example investment decisions. Three theory-based comments should be made on dynamic generality. First, intertemporal problems have no obvious impact on cooperation as pure repetition has because the Folk Theorem does not hold in general (see Dutta, 1995, and Section 2.2). Second, intertemporal spillovers increase the variety of modelling approaches significantly because every action may be made stock dependent and may itself influence several stocks in the course of time. Third, agents facing an unknown future generally expect the same situation to be repeated. Unless stationarity is too artificial an assumption for the problem at hand, it may mirror the perception of bargaining agents appropriately.

Another reservation concerns three presentation issues. First, the restrictions imposed on generality allow specific functions instead of general sets to be employed. Second, results will not be given in the form of propositions and proofs. Balancing rigidity and readability, the book will bias presentation in favour of readability and economic understanding. Third, the essentials of the theory of repeated games and strategic bargaining theory will neither be proved nor demonstrated in general. For a general exposition of both theories, the reader is referred to the references given in Chapter 2 (which are selective and not comprehensive) or to a textbook on game theory, for example, Fudenberg and Tirole (1991).

2. Strategic Bargaining and Self-enforcing Contracts

2.1 INTRODUCTION

This chapter develops the theoretical foundations for all subsequent chapters. It is organized as follows: Section 2.2 summarizes the state of the art by overviewing the literature on cooperation in a non-cooperative environment. Since both the time horizon and the discount factor are central to the approach of this book, Section 2.3 will introduce their role in economics, and it will show that the discount factor can explain more than just the pure time preferences of a single person. Section 2.4 develops the first ingredient of a theory of bargaining for self-enforcing contracts. Since repetition is a crucial feature of self-enforcing contracts, the theory of repeated games serves as a starting-point for this section. This section defines self-enforcing contracts and interprets the assumptions made for this kind of contract. Self-enforcing contracts will be defined as serving a specific strategy specification which meets the conditions of weak renegotiation-proofness according to Farrell and Maskin (1989).

Section 2.5 discusses bargaining models of which a specific strategic bargaining model will serve as the second ingredient of a theory of bargaining for self-enforcing contracts. Without claiming to cover the whole literature on bargaining, this section starts with the axiomatic approach which is due to the seminal work of John Nash (1950, 1953). It discusses the so-called Nash programme by developing the strategic bargaining approach which is basically due to Rubinstein (1982). The essence of the Nash programme is 'to test abstract or informal reasoning with simple but specific negotiation models' (Binmore and Dasgupta, 1987a: p. 9), and the strategic bargaining approach is a negotiation model which employs a simple but also fairly general bargaining rule. Several features of this approach will be discussed, especially the relevance of subgame perfection for uniqueness and the impact of time-consistent bargaining behaviour on the bargaining results in different periods. Additionally, an interpretation of the solution will be given in terms of concessions. Then, the strategic bargaining approach will be shown to be able to determine the agreement taken from the set of self-enforcing contracts in the sense of Section 2.4.

Section 2.6 points to intertemporal differences between self-enforcing and completely enforceable contracts which are likely to remain unrecognized when the following chapters focus on comparing enforceable stationary contracts with self-enforcing contracts. Section 2.7 gives an informal overview of this chapter. It collects the arguments and presents the essential results. Readers not interested in the formal development of these results may skip over Sections 2.4 and 2.5 of this chapter and may turn to the chapters on applications after reading Section 2.7.

2.2 THE STATE OF THE ART

In many cases, agreements between two parties cannot rely on external enforcement. The lack of external enforcement holds for international agreements, for example, trade and environmental agreements, for agreements which are prohibited, for example collusion among suppliers, and for agreements for which external enforcement is too costly. When enforcement by third parties is not possible, cooperation must be enforced by other means, or one must conclude that cooperation is impossible in this case.

The theory of repeated games has demonstrated that cooperation may emerge when the same issue is likely to be on the agenda again in the future. In this case, repetition may play a disciplinary role for cooperation because future behaviour may be made dependent on past actions. The Folk Theorem demonstrated that all outcomes including perfect cooperation and no cooperation may be sustained by repetition. Hence, the good news is the possibility of cooperation, and the bad news is that many outcomes are possible so that predicting the result is not possible without further assumptions.

The Folk Theorem for infinitely repeated games whose equilibrium is subgame perfect has been proved under quite different assumptions. This section will demonstrate the Folk Theorem using an example of a supergame. A supergame is defined by infinite repetition of a one-shot game (for the first papers on supergames, see Friedman, 1971; Rubinstein, 1979). Repetition means that the same situation is repeated such that after a period of unity length both agents are in the same position to decide on their strategies. Consider a game in which two players have to choose pure strategies simultaneously and in ignorance of the strategy of their opponent. They have to choose between two strategies I (called cooperation) and II (called non-cooperation). After playing the game in a certain period, the game is repeated. The example assumes that both players have almost complete information and discount future benefits. Table 2.1 gives this game in strategic form and the respective payoffs.

Table 2.1 Payoffs from cooperation and non-cooperation for two options for agent i and agent j

Agent *i* chooses rows Agent *j* chooses columns	I	II
I	(*a,a*)	(*c,b*)
II	(*b,c*)	(0,0)

Table 2.1 assumes that both agents have to decide simultaneously about their individual actions and that they take either I or II. (*i,j*) is the payoff vector and reads (payoff of agent *i*, payoff of agent *j*). Condition (2.1) ensures that the game is of the prisoners' dilemma type which makes defection from the cooperative outcome (*a,a*) profitable:

$$b > a > 0 > c. \tag{2.1}$$

In a non-cooperative setting of a one-shot game, (0,0) gives the payoffs of both agents and is obviously an unsatisfactory result as gains from coordination are left unexploited. The Folk Theorem asserts first that – given a sufficiently high discount factor – any other outcome between the non-cooperative outcome and the cooperative result (including both) can represent an equilibrium, and second that the chances for cooperation increase with the discount factor. The first effect will be shown when more than two options for each agent are discussed. The basic idea is that repetition allows for punishment as the strategies taken in future periods can be made dependent on past behaviour. One punishment strategy is the so-called trigger strategy. If each agent chooses the trigger strategy, he/she chooses cooperation as long as his/her opponent has chosen cooperation after which he/she chooses non-cooperation for the rest of the supergame.

Reverting to the non-cooperative one-shot equilibrium is credible because this strategy implies payoffs which cannot be improved upon unilaterally. Note that the trigger strategy does not allow cooperation to be resumed if one agent has cheated on his/her partner. The credibility of this threat is simply based on the observation that the one-shot outcome is an equilibrium for the supergame as well and no agent is able to improve on this outcome, given the strategy of the other agent. Suppose that both agents have the same discount factor δ and consider the example in Table 2.1. The cooperative outcome (a,a) could be sustained by a strategy which starts with cooperation and reverts to II in the case of defection of the other agent if

$$\frac{a}{1-\delta} \geq b \Leftrightarrow \delta \geq \frac{b-a}{b} \qquad (2.2)$$

holds. $a/(1-\delta)$ is the discounted sum of present and future payoffs from cooperation which must not fall short of the gains of defection, b. If δ approaches one, that is, both agents are almost perfectly patient, (2.2) is fulfilled for every a–b combination, hence for all a's which improve on (0,0). Equation (2.2) shows also that the chances for cooperation for a given a–b combination increase with the discount factor. If both agents choose the strategy not to cooperate in all periods, this result is an equilibrium in the supergame as well. This is the basic result of the Folk Theorem: for sufficiently high discount factors, all rational outcomes can be sustained as an equilibrium in the supergame. The proof of the Folk Theorem for general games has been presented by Fudenberg and Maskin (1986), Abreu, Dutta and Smith (1994) and Wen (1994). Fudenberg, Levine and Maskin (1994) demonstrate that the Folk Theorem holds even when past actions are not observable but influence a stochastic variable. All papers show that every attainable outcome can be sustained as a non-cooperative equilibrium of a supergame, and that the chances for cooperation increase with the discount factor.

The example given by Table 2.1 and the specification given by (2.2) describe the general structure of all models to be discussed in this book. Instead of the binary choice between strategies, however, an infinite number of strategy options will be assumed. The assumption of infinite repetition plays a crucial role for the Folk Theorem if the stage equilibrium is unique. Consider again the trigger strategy and suppose that repetition is finite and that both agents' strategy requires that they choose I unless the other firm has chosen II, after which an agent reverts to II for the rest of all possible realizations. Even if both agents are very patient, that is, $\delta \rightarrow 1$, finite repetition makes this strategy not credible because all punishment options are lost for the last realization. As no punishment can prevent an agent from deviating from the cooperative outcome in the last period, the non-cooperative outcome is the only equilibrium of this period. If the last period's equilibrium is the non-cooperative one-shot equilibrium, punishment options are also lost for the last but one period, thereby implying that the one-shot equilibrium gives the solution for the last but one period, too. Going back from the end of the game allows only the one-shot equilibrium to constitute an equilibrium for all periods of the game.

Benoit and Krishna (1985) have demonstrated that cooperation can be sustained in finitely repeated games if at least two one-shot Nash equilibria exist, one of which Pareto dominates the other. In this case, the agents' strategy is supposed to revert to the 'worse' equilibrium if one agent deviates from the cooperative outcome. If no agent has defected before the last period is reached, each agent's strategy is to take the 'better' equilibrium in the last period which may prevent defection in the next to last period. However, uniqueness of the one-shot equilibrium does not enable an agent to inflict damage by adopting a worse equilibrium and requires that both agents perceive the stream of possible realizations as potentially unlimited. Since this assumption is crucial, the next section will comment on its appropriateness in detail.

The Folk Theorem has been proved under different assumptions as well. Dropping the assumption of complete information, reputation can play a role in sustaining an equilibrium other than the unique one-shot Nash equilibrium in finitely repeated games (for reputation and repetition in prisoners' dilemma games, see Kreps, Milgrom, Roberts and Wilson, 1982). Bernheim and Dasgupta (1995) demonstrate that the Folk Theorem also holds in infinitely repeated games when the discount factor declines. Kandori (1992) and Smith (1992) derive a Folk Theorem for models with overlapping generations. In all models, the whole set of attainable outcomes may result from a long-run equilibrium.

The Folk Theorem holds for repeated games. Repetition is a very specific assumption with respect to the time structure of a game. In general, games involving a time structure can be dynamic so that the action space and the utilities are not stationary but change endogenously as a result of past actions. An example of the impact of past actions on the future action space is the exploitation of a non-renewable resource: the action space in the first period is limited by the available resource stock, whereas the action space in the nth period is limited by the resource stock and the actions taken in all preceding $(n - 1)$ periods. An example of changing utilities is the voluntary contribution to a public capital good: suppose that in every period each agent may contribute to an investment into public capital. If the utility of each player increases with the capital good, it is clear that each player's utility changes in the course of time.

From this perspective, repeated games are only a subset of the set of dynamic games because they make the specific assumption that neither the action space nor the utility functions change. Dutta (1995) has demonstrated that this distinction is significant because the Folk Theorem does not hold for dynamic games in general. Repeated games imply a cooperative and a non-cooperative equilibrium which do not depend on the discount factor but the equilibria of dynamic games do in general. This dependence implies that the equilibria themselves vary with the discount factor, and it implies that the chances for cooperation do not necessarily increase with the discount factor.

The ambiguous role of discount factors is due to two different effects. These effect can be demonstrated most easily by an example (for a general treatment of this problem, see Stähler and Wagner, 1998). Suppose that two firms i and j use a non-renewable resource which has to be consumed immediately after extraction. Both firms decide on resource extractions simultaneously without being able to make them dependent on the other firm's decision. The payoffs for extractions in period t can be given by exponential payoff functions:

$$\Pi_i(t) = \left[R_i(t)\right]^\gamma, \ \Pi_j(t) = \left[R_j(t)\right]^\gamma, \ 0 < \gamma < 1 \qquad (2.3)$$

where R denotes resource extractions. The resource stock and its change are given by (2.4):

$$S(t) = S(t-1) - R_i(t-1) - R_j(t-1), \ S(0) = \overline{S}. \qquad (2.4)$$

$S(t)$ denotes the stock at t. The example assumes that the game is started in period 0. The example is simple because the stock is no direct argument in the payoff function and a non-cooperative equilibrium exists which is rather simple. This non-cooperative equilibrium implies that both firms aim to deplete the resource completely in period 0 because any resource conservation of a firm would lead to an increased resource extraction of its opponent. Suppose that both firms have an equal chance in an uncoordinated exploitation race. Then, the expected resource extractions are

$$R_i'(0) = R_j'(0) = \frac{\overline{S}}{2}, \ \forall t \geq 1: \ R_i'(t) = R_j'(t) = 0. \qquad (2.5)$$

The prime denotes this non-cooperative equilibrium. The optimal solution, however, is the solution of (2.6):

$$\max_{R_i(t),R_j(t)} \sum_{t=0}^{\infty} \delta^t \left[\Pi_i(t) + \Pi_j(t)\right]$$

s.t. $\bar{S} - \sum_{t=0}^{\infty}\left[R_i(t) + R_j(t)\right] = 0,$ (2.6)

which implies the optimal resource extraction path (2.7):

$$R_i^*(t) = R_i^*(t) = \bar{S}\frac{1-\delta^{1/1-\gamma}}{2}\delta^{t/1-\gamma}.$$ (2.7)

The star denotes the optimal equilibrium path. From (2.7), one finds that the cooperative result implies a path which depends on the discount factor. Note that the non-cooperative path (2.5) and the cooperative path coincide if $\delta = 0$. Two firms which do not care about the future, also share the resource equally in a cooperative solution so that $\delta = 0$ raises no coordination problem if (2.5) gives the non-cooperative equilibrium. This result does not hold for $\delta > 0$ because cooperation improves the discounted sum of present and future payoffs. This cooperation is sustainable if unilateral defection does not pay. Suppose that unilateral defection from the cooperative path implies that a firm, say firm i, extracts the rest of the resource minus the cooperative resource extraction of the other firm, say firm j (for details of defection options, see Stähler, Wagner, 1998):

$$R_i^b(t) = S(t) - R_j^*(t).$$ (2.8)

b denotes a breach of implicit cooperation. After defection, the game is over because the resource is no longer available. Hence, defection is not profitable if the discounted sum of future cooperative profits does not fall short of the profits of defection:

$$\forall t \geq 0: \sum_{\tau=t}^{\infty}\delta^{\tau}\left[R_i^*(\tau)\right]^{\gamma} \geq \delta^t\left[R_i^b(t)\right]^{\gamma}.$$ (2.9)

Equation (2.9) must hold for all periods. Rearranging (2.9) gives a compliance condition $c[\delta,\gamma]$ which does not depend on t:

$$c(\delta,\gamma) = (1-\vartheta)^{\gamma-1} - (1+\vartheta)^{\gamma} \geq 0, \quad \vartheta = \delta^{1/1-\gamma}.$$ (2.10)

c must be non-negative for all t in order to avoid defection of a firm from cooperation. Surprisingly, c is always positive if $0 \leq \gamma \leq 0.5$. This result has been generalized by Stähler and Wagner (1997) who show that cooperation is never at risk if the elasticity of the marginal payoffs, that is, $1 - \gamma$ in this example, lies in the range [0.5, 1]. If $0.5 < \gamma < 1$, the discount factor plays a role, but an ambiguous one. When the discount factor increases, two effects can be observed: first, the future non-availability of the resource after defection is given a stronger weight. This effect makes defection less profitable. Second, the cooperative solution is changed in favour of more resource conservation. More resource conservation, however, implies that defection is made more profitable because the remaining stock in every period which may be seized unilaterally is increased. Figure 2.1 shows the results of simulating (2.10) for $\gamma = 0.7$.

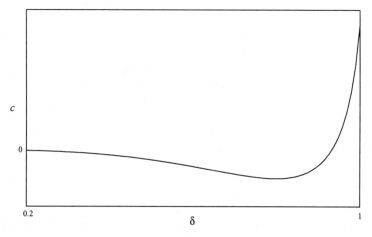

Figure 2.1   Compliance with cooperative resource extraction for $\gamma = 0.7$

From Figure 2.1, it can be seen that c falls with δ over a wide range. Hence, the second effect dominates the first effect over a wide range. This simple example reveals the ambiguous impact of the discount factor on the chances of cooperation in dynamic games. If the chances of cooperation cannot be predicted in simple dynamic games, they cannot be expected to be predicted in more complex dynamic games. This is the reason why this book concentrates on repeated games. When a strategic bargaining solution has to be developed, a model which cannot distinguish between the effect on the solution and the effect on the solution set is hard to interpret. The ambiguous role of the discount factor in dynamic games, however, should not lead to the conclusion that strategic bargaining is not capable of explaining cooperation in dynamic games. But on a theoretical level, the effects are unlikely to be so clear-cut as in repeated games. Chapter 6 will return to this point.

2.3 ON DISCOUNT FACTORS AND TIME HORIZONS

When the scope for long-run cooperation is discussed, a measure is needed which indicates the influence of future actions on current behaviour. Such a measure is the discount factor. The Folk Theorem demonstrates that the chances for cooperation in a repeated game are better the higher the discount factor is. The discount factor is at least a measure for the intertemporal preference of an agent. The interpretation of a discount factor for an individual is straightforward. A discount factor of size δ indicates that enjoying one unit utility in the next period is worth δ units utility in the current period. The higher the discount factor is, the higher are the chances for cooperation because future responses to current actions are given more weight. Since the work of Böhm-Bawerk (1912), economists discuss the role of discounting for economic behaviour. While Böhm-Bawerk characterized discounting as a psychological fact which is due to the finite horizon of human actions, more recent contributions have shown that discounting is a property of intertemporal utility functions also covering an infinite horizon (see, for example, Fishburn and Rubinstein, 1982). While Böhm-Bawerk was convinced that future events are given too little weight, recent contributions emphasize that discounting is a natural consequence of intertemporal preferences.

Throughout this book, the impatience of both agents will be given by stationary discount factors δ_i and δ_j. Let $\overline{\Pi}(t)$ denote any fixed payoff of an agent realized in t. δ_i and δ_j determine the intertemporal trade-off between realizing a unit payoff today or one period of unity length later. They measure the utility fraction of one unit utility which is left if it is realized one period later:

$$\delta^{\xi}\overline{\Pi}(\sigma) = \overline{\Pi}(\sigma+\xi), \ \sigma,(\sigma+\xi) \geq 0 \qquad (2.11)$$

The bars indicate that the current payoff is kept fixed for period σ and period $(\sigma + \xi)$. The discount factors lie in the range between 0 and 1 and δ is equal to $1/(1 + r)$ where r denotes the time preference rate. Discount factors are calibrated on a unit period. The unit period is the period between two potential realizations in a repeated game. The longer they are, the lower is the discount factor. Equation (2.11) contains a stationarity assumption because the intertemporal trade-off is measured only in terms of the period length but not in terms of periods because δ does not depend on time. This assumption is not too restrictive for an infinite time horizon. In cases of identical discount factors, this book will use this δ to indicate that $\delta_i = \delta_j (= \delta)$ holds.

The discount factor is not only an indicator of an agent's intertemporal preference. It also mirrors the intertemporal performance of organizations which are represented by agents. *Ceteris paribus*, one may expect that future responses to current actions do not play this role in organizations which they play for an individual. In an organization, several agents decide on the organization's behalf. These agents are typically at risk of being replaced. For example, the manager of a firm may be replaced when profits are too low or he/she may be able to take a better position if the firm has perfomed well in the past. Another example is a policy maker who negotiates on trade liberalization. He/she is at risk of being replaced by the next elections. Obviously, the chances of being replaced affect the intertemporal performance of an organization. When chances of replacement are high (low), future responses to current actions are strongly (weakly) taken into account because the agent is unlikely (likely) to be in the same position then. In general, whether future responses are taken strongly into account by decision makers depends on the organizational design. If there is a high risk that decision makers will lose their long-run responsibility because they are likely to be replaced in the near future, potential future responses to current actions will change current actions only negligibly. If the organizational design guarantees some continuity and decision makers are likely to carry long-run responsibility, future responses may govern current actions. The intertemporal performance of an organization can be referred to as 'institutional toughness'. The discount factor increases with institutional toughness, which signals the intertemporal performance of an organization. This discount factor results from the organizational design which determines the long-run perspectives of decision makers and hence the weight given to future responses when considering current actions. Therefore, it should also be clear that discount factors differ among agents irrespective of whether they are individuals or represent organizations.

The strategic bargaining approach as well as the theory of infinitely repeated games assume that the agents whose behaviour is explored live for ever. Obviously, this is not a realistic assumption as every human life is finite. When an agent faces a definite end of the potential negotiations, Section 2.5 will show that the bargaining solution to be developed depends on who is the first mover, who is the last mover and how long the planning horizon is. When repetition of realizations is finite, backward induction has shown that cooperation cannot be sustained if the problem under consideration has only a unique solution in the one-shot game. Thus, finite horizon assumptions imply only extreme solutions for both theories unless reputation and information asymmetries can play a role.

This book will take the position that there is no definite end of all activities but eventually a certain risk of breakdown which can be computed and integrated into the discount factors. This assumption does not indicate that breakdown risks are negligible but that the discount factors are lower the higher the risks of breakdown are felt by the agents. In general, it depends on the problem in question whether an infinite or a finite horizon approach is appropriate, especially on the perceptions of the participating agents (Rubinstein, 1992). The author's impression is that agents do not believe in definite ends, even in the case of a specified definite end as the ultimate bargaining games have demonstrated (see Güth, Schmittberger and Schwarze, 1982; Güth and Tietz, 1990; Thaler, 1988). In ultimate bargaining games, individuals were assured that no further response to their actions could occur other than those announced by the experimenter. However, the outcome was more cooperative than the theory would have predicted. Interpreting these experimental results, a specific social code to play cooperatively is often conjectured to govern decisions. But the players may also believe that the game is not definitely finished even if the rules of the game specify a definite end. Obviously, games with a definite and well-known end which is accepted by all players are rare and may be restricted to games in the literal sense.

Therefore, assuming a definite end of interactions is no weak assumption because it specifies that all agents know (and accept) the definite end of activities. If the chance of a game's breakdown is not a certain but an uncertain event, tackling this uncertainty by a risk premium allows the problem to be dealt with as a problem of potentially infinite repetition (see, for example, Fudenberg and Tirole, 1991: p. 148). Let the purely impatience-based discount factor be denoted by a tilde and assume that a constant β can mirror the probability that the game ends, for example that an agent representing an organization is replaced before the next period is reached. A constant β assumes that the risk of breakdown is exogenous so that it does not depend on the history of previous cooperation or previous negotiations. β implies an expected discounted value

$$\hat{\delta} = (1-\beta)\tilde{\delta}$$

of receiving one unit utility one period later. $\hat{\delta}$ denotes the adjusted discount factor which can be used as a certainty equivalent if all agents are risk neutral.

2.4 THE SCOPE FOR COOPERATION IN A NON-COOPERATIVE ENVIRONMENT

This section deals with the scope for long-run cooperation in a non-cooperative environment. In a non-cooperative environment, no agreement can be enforced by third parties. This section will discuss in more detail the role that repetition may play in sustaining other than the non-cooperative equilibrium. Because the credibility of strategies is central to all approaches undertaken in this book, it concentrates on subgame-perfect equilibria and on refinements introduced by the notion of renegotiation-proofness. As these refinements allow for sustaining cooperation by repetition but cannot exclude non-cooperation, the role of communication between agents will be discussed. Communication links the discussion of repeated games with the notion of self-enforcing contracts which will be defined in this section.

Several assumptions are necessary. First, both agents are assumed to have complete information about the other agent's payoff function and discount factor and the history of decisions taken in the past, and every agent knows that the other agent is informed, and the other agent knows that the one agent is informed and so on. They have no information about the actions taken by the other agent when they have to move simultaneously, but information will be revealed after both have moved simultaneously. This assumption is an almost complete information assumption so that simultaneous moves cannot be made dependent on the other agent's move. It is also a standard assumption for repeated games.

Second, two agents are considered and the utility of both depends on both agents' strategies. The two agents under consideration are called agent i and agent j. Their utility is assumed to be represented by well-defined payoff functions which depend on the agents' strategy levels s_i and s_j. In most cases, this book will assume scalars for the strategies of each agent as Chapters 3 and 5 do. In Chapter 3, the strategic efforts of each agent are the contributions to a collective good; in Chapter 5, the strategic efforts of a buyer are the transfers paid to a seller, that is, the price for a certain supply; and the strategic effort of a seller is the size of delivery to a buyer. In Chapter 4, however, transfers between agents are allowed to support the production of the collective good. In this model, one agent's policy may employ two instruments as he/she carries a certain contribution and pays transfers to the other agent (the other agent's policy involves only his/her contribution). But Chapter 4 will also demonstrate that every policy will alter maximally two out of these three instruments whereas at least one strategic variable is fixed. Hence, limiting the number of instruments to two seems not too restrictive.

Let k denote agent i or agent j and $-k$ denote the remaining agent who is not k. Equation (2.12) gives the current payoffs of both agents and states that these payoff functions do not change in the course of time. Hence, the game structure is repetition so that past actions change neither the future action space nor the payoff function. For every realization, both agents' total current payoffs are a twice differentiable function of both agents' strategy levels:

$$\forall k \in \{i, j\}: \quad \Pi_k = \Pi_k(s_i, s_j), \quad \frac{\partial^2 \Pi_k}{\partial s_k^2}(\cdot) < 0. \tag{2.12}$$

Additionally, the strategy vector $[s_i, s_j]$ is supposed to belong to a closed subset of $\mathbb{R}_+^2$. Equation (2.13) specifies that a unique strategy level exists for each agent which maximizes his/her payoff for a given strategy level of the other agent:

$$\exists s_k'(s_{-k}): \quad \frac{\partial \Pi_k}{\partial s_k}\left[s_k'(s_{-k}), s_{-k} \right] = 0. \tag{2.13}$$

Furthermore, it is assumed that only one non-cooperative Nash equilibrium exists which gives the result if the game between both agents is played only once. Equation (2.13) specifies the reaction functions s'_k. The assumption of a unique non-cooperative Nash equilibrium implies that the reaction curves intersect only once. Reaction functions will be denoted by a prime, and the payoffs realized in the non-cooperative equilibrium will be given by Π'_k. Π_k does not depend on time because identical strategies realized at different dates give both agents the same instantaneous payoff. This assumption allows impatience to be dealt with exclusively in terms of stationary discount factors.

The classical application of the theory of repeated games is cooperation between two duopolists in an industrial market (see, for example, Chapter 6 of Tirole, 1988). In a one-shot game, two duopolists arrive at a non-cooperative equilibrium in which their joint profits are not at a maximum. A basic question of the theory of repeated games was whether duopolists may improve their profits jointly when their actions are repeated. In order to exemplify the results, this section will often pick up this example. In this case, Π_k in (2.12) denotes the profits of firm k.

Since a strategic bargaining model will be employed in order to determine the distribution of bargaining gains, the model will also use net payoff functions. They are the difference between the realized payoff levels and the unique non-cooperative payoff levels:

$$\forall k \in \{i, j\}: \quad U_k(s_i, s_j) := \Pi_k(s_i, s_j) - \Pi'_k,$$

$$\exists \left[s_i^*, s_j^* \right]: \quad U_k\left(s_i^*, s_j^* \right) > 0. \tag{2.14}$$

The Nash equilibrium defines the disagreement outcome against which the potential bargaining gains are measured. It is assumed that the unique Nash equilibrium implies non-negative strategy levels. Equation (2.14) also guarantees that the game is of the prisoners' dilemma type because at least one strategy combination exists for which each firm's bargaining gains are strictly positive. For example, two firms benefit by jointly reducing their outputs or jointly increasing their prices. It should be clear that every differentiation of the net utility function (2.14) is identical with the differentiation of the payoff function (2.12) because both differ only by the constant Nash equilibrium payoffs.

The literature on repeated games assumes that the time to elapse between realizations is unity:

$$\forall \sigma \in \mathbb{N}_0 : r(\sigma) = \sigma. \qquad (2.15)$$

In (2.15), r denotes the number σ realization, and (2.15) implies that the dating of realizations is fixed. This assumption will be relaxed in the next section when a strategic bargaining model is employed, and it will be shown that realizations will in fact take place as soon as possible and that the outcome but not the options meet the time structure of (2.15). Even now, however, it should be carefully noted that realizations have a completely different time structure as bargaining because bargaining will be assumed to be based on costless communication, whereas chances for realizing bargaining gains are rare because they are repeated only after a certain time.

All arguments in this section will be developed along two lines of reasoning: first, all conditions for long-run cooperation will be given by using the payoff functions (2.12). These payoff functions may imply an infinite number of solutions which will fulfil the conditions for long-run cooperation. Second, the example given by Table 2.1 in Section 2.2 can be understood as mirroring a limited number of possible outcomes, and this binary choice model will support the intuitive understanding. This example should not be understood as an alternative approach but as an example for the general payoff functions from which only a limited number of possible outcomes has entered Table 2.1. Contrary to the general conditions, this example assumes that both firms have the same discount factor.

Trigger Strategies

If two firms face definitely only one realization, the non-cooperative Nash solution is the outcome of their interactions unless they can themselves credibly be bound by a contract. A credibly binding contract ensures that any deviation from the terms of the contract is either effectively punished by a third party or a third party is able to enforce the terms of the contract. Obviously, enforceable contracts are not necessary if the problem under consideration offers no benefits from cheating because the non-cooperative Nash equilibrium cannot be improved upon by another action pair taken. However, if the problems under consideration are of the prisoners' dilemma type, the so-called cooperative outcome is no equilibrium in a one-shot game as every firm can improve on its utility by defection. Table 2.1 gave a simple example for this situation, whose general structure was assumed by (2.14). If the Nash equilibrium induced the cooperative outcome instead, no problem of bargaining could arise as no bargaining gains are to be realized by coordination.

Let the effort levels which are sustained by repetition be denoted by '^' in the remainder of this section. The trigger strategy is successful if the gains from infinite cooperation exceed the gains from defection and no cooperation in all following periods. This condition is given in (2.16):

$$\forall k \in \{i,j\}: \frac{1}{1-\delta_k} U_k\left[\hat{s}_k, \hat{s}_{-k}\right] - U_k\left[s'_k\left(\hat{s}_{-k}\right), \hat{s}_{-k}\right] \geq 0. \quad (2.16)$$

The first term in (2.16) gives the discounted sum of present and future bargaining gains. The second term gives the gains from unilateral defection. Recall that the bargaining gains in the non-cooperative one-shot equilibrium are zero.

At least two arguments have cast doubts on the reliability of the trigger strategy to sustain cooperation. One argument, first raised by Abreu (1988), was that reverting to the non-cooperative outcome may not suffice to support the cooperative outcome (insert for example $a = 3$, $b = 8$, and $\delta = 4/7$ into (2.2)). Abreu discussed penal codes which can replace reverting to the non-cooperative outcome. These penal codes specify a certain action profile to be taken in the case of defection of a firm, including defection from the penal code. Abreu demonstrated that penal codes may sustain more cooperation than the simple trigger strategy. But once a firm has defected, there is no way back to the original outcome.

Another argument which has received increasing attention in the literature concerns the credibility of the trigger strategy. The trigger strategy obviously assumes cooperation in the beginning but rules out any return to cooperation after one firm has defected. Roughly speaking, it sounds a little bit counterintuitive that two firms choose strategies which start with cooperation but revert to non-cooperation if one incidence of deviation occurs. The literature on renegotiation-proofness deals with the option that firms may reconsider their strategies and talk about revising their original plans after a firm has chosen deviating behaviour. Renegotiation-proofness refines the equilibrium outcomes by requiring that any outcome should be immune against renegotiations. The shaping papers of this strand of literature are

Abreu, Pearce and Stacchetti (1989), Bernheim and Ray (1989), Farrell and Maskin (1989) and van Damme (1989), but definitions given in these papers are quite different.

Weak Renegotiation-proofness

Before going into the details of renegotiation-proofness which fits the notion of self-enforcing contracts, a short discussion on the credibility of threats is helpful. Credibility of different strategies can be discussed by asking the question how strongly past events influence future behaviour (see Farrell and Maskin, 1989; Mohr, 1988). On the one hand, only the one-shot equilibrium can constitute an equilibrium in the supergame if firms' decisions are supposed to be highly history independent such that past events do not influence future behaviour. This strategy specification is called Markov perfect and is similar to the sunk cost argument that bygones are bygones (Maskin and Tirole, 1994). However, this similarity does not make only Markov-perfect strategies convincing because the relevance of the past for future outcomes and the reaction to past behaviour should not be confused. If strategies do not depend on past events, cooperation is impossible. On the other hand, cooperation is possible if a firm's strategy does depend strongly on past events because it denies any cooperation if the other firm has defected in the past. In this case, a firm does not forgive defection even if defection occurred, say, fifty years ago.

The concept of renegotiation-proofness lies in between these extreme approaches. As Bernheim and Ray (1989) have put it, an equilibrium should not 'prescribe any course of action taken in any subgame that players would jointly wish to renegotiate (given the restriction that any alternative must themselves be invulnerable to subsequent renegotiation)' (p. 297). The strategies which are renegotiation proof are also history dependent because punishment follows deviating behaviour, but they also allow both firms to resume cooperation. The option to resume cooperation makes such strategies immune against renegotiations (that is, renegotiation proof) because these strategies do not require the realization of Pareto-inferior non-cooperative payoffs in all future periods as the trigger strategy does. It is obvious that this compromise between the need for punishing deviating behaviour and the resumption of cooperation potentially restricts the scope for cooperation further compared to pure punishment strategies because a potentially deviating firm anticipates resuming cooperation. Therefore, the punishment threat may be weakened and renegotiation-proofness may impose stricter restrictions on sustainable cooperation.

This book follows the concept of weak renegotiation-proofness according to Farrell and Maskin (1989). Their approach is purely non-cooperative and mirrors renegotiation-proofness in a non-cooperative environment (for a discussion of renegotiation in repeated games, see Abreu and Pearce, 1991; Pearce, 1992). It not only discusses cooperation on the basis of weak renegotiation-proofness but on the basis of strong renegotiation-proofness as well. This section will turn to strong renegotiation-proofness after having developed weak renegotiation-proofness.

According to Farrell and Maskin, an equilibrium is weakly renegotiation proof if none of its continuation payoffs is Pareto dominated by another continuation payoff. This kind of renegotiation-proofness is weak because it does not consider substitution of punishment by another agreement (which defines strong renegotiation-proofness). Continuation payoffs are the sum of all discounted present and future payoffs. Pareto dominance requires that these discounted values of payoffs which the strategies give both firms after defection of a firm has occurred should not be Pareto dominated by another strategy specification. A strategy which specifies that the non-deviating firm reverts to the non-cooperative outcome as long as the deviating firm has chosen cooperative behaviour unilaterally in one period can be renegotiation proof (see van Damme, 1989). More generally, weak renegotiation-proofness may be ensured by the strategy specification that the non-deviating firm reverts to the non-cooperative outcome as long as the deviating firm has chosen cooperative behaviour unilaterally in n periods.

Compared to the trigger strategy, the deviating firm is better off if an investment in a resumption of cooperation pays, and the non-deviating firm is better off as it enjoys one free ride (or n free rides) and a resumption of cooperation if the other firm wants to resume cooperation. In this case, the continuation payoffs Pareto dominate the payoffs of reverting to the one-shot equilibrium for ever and are not Pareto dominated by returning to the original cooperation without punishment. Note that the approach adopted here is not restricted to Pareto optimal outcomes, whose optimality results from cooperative behaviour. Instead, feasibility restrictions imposed by weak renegotiation-proofness may rule out Pareto-optimal outcomes but will allow the realization of outcomes which Pareto dominate the one-shot Nash equilibrium. For sufficiently high discount factors, the feasibility of Pareto-optimal outcomes was proved by Evans and Maskin (1989).

Obviously, weak renegotiation-proofness should also meet conditions (2.2) or (2.16), respectively. If these conditions did not hold, one firm would deliberately defect and refuse to cooperate in the future. Additionally, two conditions have to be met which will be referred to as *ex ante* and *ex post* compliance. *Ex ante* compliance requires that it must not pay for a firm to defect in one period and to invest in a resumption of cooperation in n periods compared to cooperation in these $n + 1$ periods. This condition is given by (2.17).

$$\forall k \in \{i,j\}: \frac{1-\delta_k^{n+1}}{1-\delta_k} U_k\left[\hat{s}_k, \hat{s}_{-k}\right] - U_k\left[s_k'\left(\hat{s}_{-k}\right), \hat{s}_{-k}\right]$$

$$-\delta_k \frac{1-\delta_k^n}{1-\delta_k} U_k\left[\hat{s}_k, s_{-k}'\left(\hat{s}_k\right)\right] \geq 0. \tag{2.17}$$

The first term gives the discounted sum of bargaining gains of cooperation in $(n+1)$ periods. $(1-\delta_k^{n+1})/(1-\delta_k)$ is the factor giving the sum of a finite geometric series which covers $(n+1)$ periods. The second term gives the defection benefits in the period of defection. The third term gives the (negative) benefits from being punished in n periods in order to resume cooperation. $\delta_k(1-\delta_k^n)/(1-\delta_k)$ is the factor giving the sum of a finite geometric series which covers n periods from the second period on. For the example specified by Table 2.1 and the assumption that the punishment length is restricted to one period, *ex ante* compliance requires (2.18):

$$(1+\delta)a \geq b + \delta c \Leftrightarrow \delta \geq \frac{b-a}{a-c}. \tag{2.18}$$

Ex ante compliance does not suffice to guarantee weak renegotiation-proofness because a deviating firm must prefer to resume cooperation after it has enjoyed defection benefits. This condition is called *ex post* compliance:

$$\forall k \in \{i,j\}: \quad \frac{1-\delta_k^n}{1-\delta_k} U_k\left[\hat{s}_k, s_{-k}'\left(\hat{s}_k\right)\right] + \frac{\delta_k^n}{1-\delta_k} U_k\left[\hat{s}_k, \hat{s}_{-k}\right] \geq 0. \tag{2.19}$$

Condition (2.19) requires that the (negative) benefits from investing in a resumption of cooperation, that is, being punished in n periods, and the discounted value of cooperation resumed after n punishment periods must not fall short of refraining from resuming cooperation. $(1-\delta_k^n)/(1-\delta_k)$ is the factor giving the sum of a finite geometric series which covers n periods, $\delta_k^n/(1-\delta_k)$ is the factor giving the sum of an infinite geometric series which starts after n periods. For the example specified by Table 2.1 and the assumption that the punishment length is restricted to one period, *ex post* compliance requires:

$$c + \frac{\delta a}{1-\delta} \geq 0 \Leftrightarrow \delta \geq \frac{-c}{a-c}. \tag{2.20}$$

Additionally, the firm which was the victim of defection (say firm $-k$) must benefit from this strategy specification compared to returning to cooperation without punishment:

$$\forall - k \in \{i, j\}: \frac{1-\delta^n_{-k}}{1-\delta_{-k}} U_{-k}\left[\hat{s}_k, s'_{-k}(\hat{s}_k)\right]$$

$$+ \frac{\delta^n_{-k}}{1-\delta_{-k}} U_{-k}\left[\hat{s}_k, \hat{s}_{-k}\right] \geq \frac{1}{1-\delta_{-k}} U_{-k}\left[\hat{s}_k, \hat{s}_{-k}\right]. \tag{2.21}$$

As the non-deviating firm enjoys n free rides after defection, its preference is obvious and (2.21) imposes no additional restriction but is fulfilled in every case. The victim of defection is always better off compared to reverting to the non-cooperative equilibrium because he/she realizes high bargaining gains during the punishment phase and cooperative bargaining gains further on. Condition (2.21) may therefore be neglected. Additionally, it can be shown easily that (2.16) can be neglected as well because it follows from (2.17) and (2.19). Multiplying (2.19) by δ_k and adding (2.21) to it leads exactly to (2.16). Similarly, multiplying (2.20) by δ and adding (2.18) to it leads exactly to (2.2). Hence, (2.17) and (2.19) or (2.18) and (2.20), respectively, are the relevant restrictions. For the example of Table 2.1, only if δ meets these two conditions, is (a,a) a weakly renegotiation-proof equilibrium:

$$\delta \geq \max\left\{\frac{b-a}{a-c}, \frac{-c}{a-c}\right\}. \tag{2.22}$$

The refinement of weak renegotiation-proofness fits the modelling framework of this book because bargaining should obviously include the possibility of renegotiation. If two firms can communicate and make proposals every time they want to, switching to non-cooperation for ever is not a credible option. However, weak renegotiation-proofness is only a necessary condition to be imposed on self-enforcing contracts because the one-shot equilibrium is of course also weakly renegotiation proof but it will not qualify for an agreed-upon self-enforcing contract. The concept of weak renegotiation-proofness does not guarantee a cooperative outcome but adds cooperative outcomes to the set of possible equilibria if the compliance constraints are fulfilled. All concepts carry the implicit assumption that two firms would agree on a Pareto-dominant outcome, but cannot rule out that the one-shot equilibrium gives the solution.

Compliance Constraints and the Solution Set

One may now consider the compliance constraints in further detail in order to discuss how the compliance constraints change the solution set. Let *CC* denote a compliance constraint in which all terms are on the LHS of the inequality sign and which is equal to zero. It is obvious that all compliance constraints are equal to zero for $U_k[\hat{s}_i, \hat{s}_j] = 0$. Hence all compliance constraints start at the non-cooperative levels since the payoffs, the defection benefits and the punishment benefits coincide at this point. The subscript of *CC* denotes the firm under consideration, the superscript denotes the respective compliance constraint: 1 denotes constraint (2.16), 2 denotes *ex ante* compliance, and 3 denotes *ex post* compliance. Condition (2.16) need not be discussed in detail, and (2.17) and (2.19) may be simplified and the slope of the compliance constraints in the $s_i - s_j$ space can be computed:

$$CC_k^2(\hat{s}_k, \hat{s}_{-k}) := \left[1 - \delta_k^{n+1}\right]U_k\left[\hat{s}_k, \hat{s}_{-k}\right]$$
$$-\left[1 - \delta_k\right]U_k\left[s_k'(\hat{s}_{-k}), \hat{s}_{-k}\right] - \delta_k\left[1 - \delta_k^n\right]U_k\left[\hat{s}_k, s_{-k}'(\hat{s}_k)\right] \geq 0$$

$$\left[\frac{ds_{-k}(\cdot)}{ds_k}\right]_{CC_k^2 = 0} =$$

$$-\frac{\dfrac{\partial U_k}{\partial s_k}\left[\hat{s}_k, \hat{s}_{-k}\right] - \dfrac{\delta_k\left[1 - \delta_k^n\right]}{1 - \delta_k^{n+1}}\left[\dfrac{\partial U_k}{\partial s_k}\left[\hat{s}_k, s_{-k}'(\hat{s}_k)\right] + \dfrac{\partial U_k}{\partial s_{-k}}\left[\hat{s}_k, s_{-k}'(\hat{s}_k)\right]\dfrac{ds_{-k}'}{ds_k}\right]}{\dfrac{\partial U_k}{\partial s_{-k}}\left[\hat{s}_k, \hat{s}_{-k}\right] - \dfrac{1 - \delta_k}{1 - \delta_k^{n+1}}\dfrac{\partial U_k}{\partial s_{-k}}\left[s_k'(\hat{s}_{-k}), \hat{s}_{-k}\right]}$$

(2.23)

$$CC_k^3(\hat{s}_k, \hat{s}_{-k}) := \left[1 - \delta_k^n\right]U_k\left[\hat{s}_k, s_{-k}'(\hat{s}_k)\right] + \delta_k^n U_k\left[\hat{s}_k, \hat{s}_{-k}\right] \geq 0$$

$$\left[\frac{ds_{-k}(\cdot)}{ds_k}\right]_{CC_k^3 = 0} =$$

$$-\frac{\dfrac{\partial U_k}{\partial s_k}\left[\hat{s}_k, \hat{s}_{-k}\right] + \dfrac{1 - \delta_k^n}{\delta_k^n}\left[\dfrac{\partial U_k}{\partial s_k}\left[\hat{s}_k, s_{-k}'(\hat{s}_k)\right] + \dfrac{\partial U_k}{\partial s_{-k}}\left[\hat{s}_k, s_{-k}'(\hat{s}_k)\right]\dfrac{ds_{-k}'}{ds_k}\right]}{\dfrac{\partial U_k}{\partial s_{-k}}\left[\hat{s}_k, \hat{s}_{-k}\right]}$$

(2.24)

One cannot deduce the slope of (2.23) or (2.24) in general and therefore which compliance constraint is dominating cannot be determined.

Additionally, one cannot derive an unambiguous sign of the second derivatives for (2.23) and (2.24) which were a necessary condition in order to be able to determine how the solution set is changed. Hence, this section assumes that the compliance constraints determine a concave and at least piecewise differentiable payoff frontier or a single Pareto-dominant point. The payoff frontier is defined by payoff combinations which are not Pareto dominated. In order to distinguish this frontier from the Pareto frontier, it will be referred to as the set of efficient agreements. Diagrams I to III in Figure 2.2 show frontiers which are allowed. Diagram IV shows a frontier which is ruled out by assumption. The dotted line gives the Pareto frontier which is not subject to compliance constraints, and the broken line gives the frontier determined by the compliance constraints which is Pareto-dominated by efficient agreements.

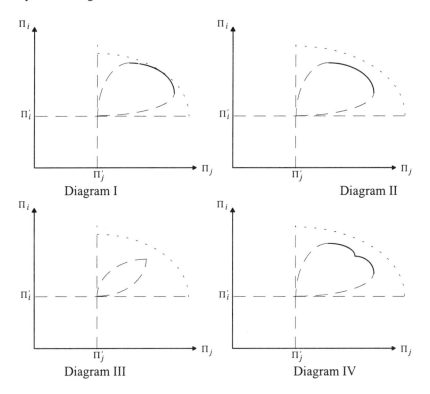

Figure 2.2   Set of efficient agreements

The set of efficient agreements is given by the frontier for which $d\Pi_i/d\Pi_j \leq$ 0 or by a single point as in Diagram III which Pareto dominates all other feasible payoff combinations. Note that it is not necessary that the whole set is convex but only that the frontier is concave and piecewise differentiable.

Although conditions (2.23) and (2.24) cannot guarantee a concave frontier, some special cases and special results can be developed. Orthogonal reaction functions allow (2.23) and (2.24) to be simplified substantially because cross-derivatives of the profit function and the slope of the reaction curves are zero and the first derivatives of the profit function with respect to one agent's strategy level do not depend on the other agent's strategy level. For orthogonal reaction functions, the slopes of the compliance constraints are given by (2.25) and (2.26).

$$\left[\frac{ds_{-k}(\cdot)}{ds_k}\right]_{CC_k^2=0} = -\frac{1}{\delta_k}\frac{1-\delta_k}{1-\delta_k^n}\frac{\partial U_k(\cdot)/\partial s_k}{\partial U_k(\cdot)/\partial s_{-k}}. \tag{2.25}$$

$$\left[\frac{ds_{-k}(\cdot)}{ds_k}\right]_{CC_k^3=0} = -\frac{1}{\delta_k^n}\frac{\partial U_k(\cdot)/\partial s_k}{\partial U_k(\cdot)/\partial s_{-k}}. \tag{2.26}$$

For $n = 1$, all slopes coincide. Since all compliance constraints start at the non-cooperative equilibrium, $n = 1$ and orthogonality imply that all compliance constraints coincide. This is the reason why Chapters 3, 4 and 5 employ models with orthogonal reaction functions because this assumption will simplify the development of the main results significantly.

Furthermore, one may consider the influence of n and δ_k on the scope for cooperation left possible by the compliance constraints. n is determined to maximize the scope for cooperation, and therefore the impact of n on the set of efficient agreements must be explored. The scope for cooperation can be measured by the critical U_k in every compliance constraint. U_k measures the bargaining gains which are necessary to make firm k indifferent between compliance and defection. If U_k is high (low), the scope for cooperation is low (high) because firm k's compliance is guaranteed by offering this firm high (low) bargaining gains. Let the critical bargaining gains be denoted by U_k^1, U_k^2 and U_k^3 for the respective compliance constraints:

$$U_k^1(\tilde{n},\delta_k) := (1-\delta_k)U_k', \tag{2.27}$$

$$U_k^2(\tilde{n},\delta_k) := \frac{1-\delta_k}{1-\delta_k^{\tilde{n}+1}}U_k' + \frac{\delta_k\left(1-\delta_k^{\tilde{n}}\right)}{1-\delta_k^{\tilde{n}+1}}U_k^p, \tag{2.28}$$

$$U_k^3(\tilde{n}, \delta_k) := -\frac{1 - \delta_k^{\tilde{n}}}{\delta_k^{\tilde{n}}} U_k^p. \qquad (2.29)$$

Equations (2.27), (2.28) and (2.29) determine the critical bargaining gains as functions of $\tilde{n}$ and δ_k. $\tilde{n}$ denotes the punishment period and replaces n which is an integer. $\tilde{n}$ is assumed to belong to the set of real numbers in order to determine the impact of changes of n by differentiation. In (2.27), (2.28) and (2.29), $[\hat{s}_i, \hat{s}_j]$ is fixed. Differentiations yield

$$\frac{\partial U_k^1}{\partial \tilde{n}}(\cdot) = 0, \quad \frac{\partial U_k^1}{\partial \delta_k}(\cdot) = -U_k' < 0, \qquad (2.30)$$

$$\frac{\partial U_k^2}{\partial \tilde{n}}(\cdot) = \frac{\ln \delta_k \delta_k^{\tilde{n}+1}(1 - \delta_k)}{\left[1 - \delta_k^{\tilde{n}+1}\right]^2} \left[U_k' - U_k^p\right] < 0,$$

$$\frac{\partial U_k^2}{\partial \delta_k}(\cdot) = -\frac{1 - (\tilde{n}+1)\delta_k^{\tilde{n}} + \tilde{n}\delta_k^{\tilde{n}+1}}{\left[1 - \delta_k^{\tilde{n}+1}\right]^2} \left[U_k' - U_k^p\right] < 0. \qquad (2.31)$$

$\partial U_k^2 / \partial \delta_k$ is negative because $1 - (\tilde{n}+1)\delta_k^{\tilde{n}} + \tilde{n}\delta_k^{\tilde{n}+1}$ has a minimum at

$$\tilde{n} = -\frac{1}{\ln \delta_k}\left[1 + \frac{\ln \delta_k}{1 - \delta_k}\right] < 0.$$

The negative sign of the minimum is due to the fundamental property of logarithmic functions that $\ln \delta_k < \delta_k - 1$ for all $\delta_k \neq 1$ (if $\delta_k = 1$, $\ln \delta_k = \delta_k - 1 = 0$). Then, the lowest feasible value for $1 - (\tilde{n}+1)\delta_k^{\tilde{n}} + \tilde{n}\delta_k^{\tilde{n}+1}$ is given by $\tilde{n} = 1$ for which

$$1 - (\tilde{n}+1)\delta_k^{\tilde{n}} + \tilde{n}\delta_k^{\tilde{n}+1} = (1 - \delta_k)^2 > 0$$

holds. As $1 - (\tilde{n}+1)\delta_k^{\tilde{n}} + \tilde{n}\delta_k^{\tilde{n}+1}$ is always positive, $\partial U_k^2 / \partial \delta_k$ is always negative. Differentiation of (2.29) yields

$$\frac{\partial U_k^3}{\partial \tilde{n}}(\cdot) = \frac{\ln \delta_k}{\delta_k^{\tilde{n}}} U_k^P > 0, \quad \frac{\partial U_k^3}{\partial \delta_k}(\cdot) = \tilde{n} \delta_k^{-(\tilde{n}+1)} U_k^P < 0 \qquad (2.32)$$

The negative signs of the partial derivatives with respect to δ_k in (2.30), (2.31) and (2.32) demonstrate the Folk Theorem for weak renegotiation-proofness since the bargaining gains necessary to guarantee compliance decrease with the discount factor. Whereas CC_k^1 does not depend on n, (2.31) and (2.32) show that an increase of n decreases the bargaining gains necessary to guarantee *ex ante* compliance but increases the bargaining gains necessary to guarantee *ex post* compliance. Obviously, an increase in the number of punishment periods makes a firm refrain from defection but, after defection has occurred, a long punishment phase must be compensated by substantial bargaining gains in order to make investment in a resumption of cooperation profitable.

n is determined in order to maximize the scope for potential cooperation. The models which the following chapters employ will imply an unambiguous n which maximizes the scope for cooperation. In general, however, different n's may have to be employed: some payoff combinations may be sustained only by a longer punishment period whereas other payoff combinations may be sustained only by a short punishment period. In the first case, *ex post* compliance is less demanding than *ex ante* compliance; in the second case, *ex ante* compliance is less demanding than *ex post* compliance. For all cases, irrespective of whether n is fixed or is changed, it is relevant that the set of efficient agreements defines a concave, at least piecewise differentiable frontier.

The Very Nature of Contracts

When several equilibria exist, one faces an equilibrium selection problem, and in order to be able to predict strategies, further assumptions have to be made. One approach is the theory of equilibrium selection, and the main contributions to this theory are Güth and Kalkofen (1989) and Harsanyi and Selten (1988). These models identify the selected equilibrium by proposing additional criteria which an equilibrium has to fulfil. These criteria are different from refinements like weak renegotiation-proofness, which define certain credibility conditions. Instead, these criteria are general and determine which one out of the multiple equilibria is supposed to be selected. These models do not ignore the fact that agents communicate but they assume that communication does not change the behaviour of agents since it does not change the payoff structure. Therefore, communication is not considered in this strand of literature.

However, it makes a difference if communication leads to an agreement which qualifies for consistent declarations of intent. Consistent declarations of intent specify an agreement so that both agents deliberately announce intentions which do not contradict each other. In this case, mutual acceptance of an outcome which implies mutual improvement has another quality even if this outcome must be sustained by repetition. The result of this process will be called a contract. Roughly speaking, a contract is the result of communication which has led to a plan which lets both agents know what to do in all imaginable circumstances.

Contracts specify strategies because both firms communicate before realizations take place, and agree upon certain strategies for all conceivable cases. The basic idea is that talking constructively about mutual improvements leads to a mutually agreed-upon contract which coordinates the behaviour of both firms. This is no cheap talk in the usual sense which reveals private information (Farrell and Gibbons, 1989) but a coordination of plans. Under almost complete information, every contract is complete. Talking about mutual improvements is given an analytical structure through a well-defined bargaining procedure when the selection of a contract is discussed. In general, a contract should meet the following specification: a contract specifies consistent declarations of intent of two agents. It contains two paragraphs which serve for pure strategy specification of both contractors. The first paragraph specifies the agreed-upon inputs of both agents; the second paragraph specifies the actions to be taken in the case of any deviation from the agreed-upon inputs specified by the first paragraph.

This definition is similar to the contract definition of civil law. Note that the form of the contracts is not specified by this definition. A contract can be based on an oral or written agreement, can be made by telephone or attested by a notary. The first paragraph specifies the strategies agreed upon, the second paragraph specifies the strategies in the case of defection which benefit one agent compared to no cooperation or no punishment (for the relevance of alternative proposals, see the discussion on strong renegotiation-proofness, below). The definition of a contract does not imply that only efficient contracts are agreed upon, but talking constructively can be expected to lead to a certain strategy specification which leaves no mutual improvement unexploited. The strategic bargaining model which the next section employs will lead to this result. Note that any assumption that contracts would lead to efficient outcomes would only mitigate the ambiguity of results because there may exist several efficient outcomes.

The set of self-enforcing contracts is a subset of the set of all contracts. The literature uses many essentially different notions of self-enforcing contracts or self-enforcing agreements. In environmental economics, self-enforcing agreements cover international agreements on international environmental problems. They have quite a different meaning, because cooperation is modelled in a one-shot framework which deals with the problem of participation (see Barrett, 1991, 1992; Carraro and Siniscalco, 1993; Heal, 1992). Self-enforcing environmental agreements are based on voluntary participation, and they are in equilibrium if no participant wants to leave the agreement and no outsider wants to enter the agreement. In labour economics, self-enforcing contracts are contracts which do not specify the behaviour of employees for all events of an unknown future (see, for example, Carmichael, 1989). Optimal contracts are the result of balancing the costs of explicit enforcement and potential breach costs (Klein, 1985). These contracts are a specific variant of incomplete contracts (Hart and Moore, 1988) and provide a link to a quite different notion of renegotiation-proofness which is applied to contracts and which should not be confused with the notion used here. If contracts are incomplete and contain specific terms on which agents could improve upon in the case of a certain event, renegotiation will be anticipated when designing the contract. Roughly speaking, a contract is renegotiation proof if its terms take into account the potential mutual improvements after new information has been revealed (see, for example, Green and Laffont, 1992; Rubinstein and Wolinsky, 1992; Dewatripont and Maskin, 1995). Complete information, however, leaves no scope for incomplete contracts.

The definition for self-enforcing contracts given here is similar to the one given by Telser (1980) who assumes that no third party is able to intervene because third party intervention is either impossible or too costly. However, Telser's defection penalty is stronger because he assumes that a firm stops the sequence of transactions in the case of defection. This book assumes that a self-enforcing contract $[\hat{s}_i, \hat{s}_j]$ is a contract which specifies an outcome which fulfils (2.17) and (2.19). The second paragraph of a self-enforcing contract specifies the actions to be taken by both contractors in the case of deviating behaviour of one contractor but no other parties than the contractors react to deviation. The second paragraph specifies the punishment plan: the non-deviating firm switches to the defection input level as long as the deviant firm has provided for its originally agreed-upon input level unilaterally in n periods.

The basic difference between an enforceable and a self-enforcing contract is to be found in the second paragraph. A self-enforcing contract must stabilize its outcome by its own means, and these means are given by a weakly renegotiation-proof punishment plan. This is the reason why these contracts are called self-enforcing. Note that both firms voluntarily agree upon this plan and do not decide on their strategies independently. Independent strategy specification does not qualify for consistent declarations of intent. Their agreement also specifies n, the number of punishment periods. *n* will be determined to balance the strength of the constraints (2.17) and (2.19). *n* will be determined in order to maximize the scope for cooperation, and as Chapters 3, 4 and 5 employ orthogonal reaction functions, it should be clear that $n = 1$ maximizes the scope for cooperation for these models. As the set of self-enforcing contracts is not necessarily a singleton, that is, more than one strategy combination defines a self-enforcing contract, it should also be clear that there is a problem of which self-enforcing contract to choose. This determination problem will be coped with using the strategic bargaining approach.

Strong Renegotiation-proofness and Rebargaining-proofness

Weak renegotiation-proofness requires that neither agent wants to renegotiate on the punishment scheme. This requirement has defined self-enforcing contracts. In addition to this requirement, strong renegotiation-proofness questions the credibility of punishment schemes when alternative agreements are a substitute for punishment. Strong renegotiation-proofness would add additional constraints to those already introduced by weak renegotiation-proofness. According to Farrell and Maskin (1989), an outcome is strongly renegotiation proof if its continuation payoffs are not Pareto dominated by another weakly renegotiation-proof agreement. The idea is that an agent who has defected is able to make a new proposal which should be a substitute for punishment and a resumption of cooperation. If the outcome of the new agreement which is proposed by the agent gives higher discounted utilities to both agents than punishment and resuming cooperation, they can be expected to refrain from the old agreement's punishment plan and consent to the new agreement. The fatal implication of refraining from punishment is that punishment loses its credibility because a potentially deviating agent anticipates that he/she will not be punished. If every punishment plan of a weakly renegotiation proof contract is Pareto dominated by another weakly renegotiation-proof contract, no strongly renegotiation-proof agreement exists (Farrell and Maskin, 1989; Schultz, 1994). This case cannot be ruled out and would imply that cooperation breaks down completely unless both agents can credibly commit themselves not to accept any proposal which is a substitute for punishment.

However, the concept of strong renegotiation-proofness entails a strong assumption when a bargaining procedure is taken into account. The strong assumption made by this concept is that an alternative agreement for all future periods has been proposed. Suppose that similar bargaining powers of both agents have led to payoffs of (5,5). If one agent has deviated from (5,5), both agents discuss (1,9) and (0,10) or (9,1) and (10,0), respectively, as a substitute for punishment and a resumption of cooperation (5,5). But agreeing upon (1,9) or (0,10) (or (9,1) or (10,0)) means specifying an agreement for all future periods which underestimates the deviating agent's, and overestimates the other agent's, bargaining power. This is a strong assumption as renegotiation would suppress all bargaining options in future periods. Suppressing all future bargaining options may also be seen as not credible if the selection of the self-enforcing contract should be bargaining based. For all future periods, the deviating agent is assumed to be unable to exploit his/her bargaining power.

The implications of strong renegotiation-proofness are basically due to the ambiguity of equilibrium selection. If no bargaining procedure determines the outcome, all self-enforcing contracts are implicitly assumed to have the same chance of being selected. In this setting, the division of bargaining gains is indeterminate and self-enforcing contracts dividing bargaining gains differently between both agents are not subject to strategic interaction. This assumption is inappropriate when bargaining power should determine the selection of the self-enforcing contract. An alternative is to consider rebargaining.

Rebargaining allows for two options. First, it allows for bargaining in the punishment phases so that mutual improvements by another weakly renegotiation-proof contract may be exploited. Suppose again that two agents have the same bargaining power and that the problem under consideration should give them identical bargaining gains. Agent i has deviated from the self-enforcing contract (5,5) and punishment gives (−2,8). Assume again that another self-enforcing contract exists which specifies (0,10). Obviously, both agents prefer this contract instead of punishment. But rebargaining requires that accepting this contract must result from a bargaining procedure as well. For this bargaining procedure, the punishment utilities serve as the reservation utilities. (0,10) gives both agents the same bargaining gains compared to (−2,8), that is, bargaining gains of two units, and (0,10) is therefore the rebargaining solution which replaces the old punishment plan. A self-enforcing contract (1,9) would not be accepted by agent j as an alternative to (0,10) because rebargaining gains were distributed which were not in line with his/her bargaining power (i's rebargaining gains were 3, j's rebargaining gains were 1).

Second, rebargaining-proofness restricts to the punishment phases the periods which are open for rebargaining. If an agent has deviated from the agreement, the punishment plan specifies payoffs during punishment which both agents may be able to improve upon. If punishment is complete, the reservation utilities are no longer given by the punishment plan but by the very definition of the problem. This is the weakest assumption to be made with respect to bargaining as defection merely shifts the reservation utilities to a level for the punishment phase determined by the punishment plan.

Then, a contract does not suffer from rebargaining if rebargaining does not destroy *ex ante* compliance but both agents are still better off by fulfilling the contract than by breaking it.

Neither strong renegotiation-proofness nor rebargaining-proofness are easy to deal with even in a model which employs specific functions. The appendix contains the respective definitions of strong renegotiation-proofness and of rebargaining-proofness. In the appendix, an example shows that rebargaining-proofness allows more cooperative outcomes which are weakly renegotiation proof to be sustained although it is not a redundant condition. The high degree of complexity associated with rebargaining-proofness would suggest developing the strategic bargaining solution for self-enforcing contracts as a first step, and checking rebargaining-proofness as a second step.

2.5 A THEORY OF STRATEGIC BARGAINING FOR SELF-ENFORCING CONTRACTS

This section develops the strategic bargaining solution for self-enforcing contracts. Before developing this solution, it is helpful to introduce bargaining theory. Therefore, this section starts with the axiomatic bargaining approach. The strategic bargaining approach will be introduced by discussing an alternating offer game with finite bargaining stage number. This discussion will demonstrate the relevance and appropriateness of subgame perfection for strategic bargaining theory. Exploring the limit of this bargaining game for an infinite horizon gives the main result of strategic bargaining theory. Both the discussion of the axiomatic approach and the basic strategic bargaining approach will assume bargaining for only one realization (which may be delayed).

Then, this section turns to bargaining for long-run cooperation. Strategic bargaining for long-term cooperation in a non-cooperative environment assumes that every agent is selfish and may submit a proposal whenever he/she wants to. The option to submit proposals which may successfully replace existing agreements specifies that bargaining behaviour for long-term cooperation should be time consistent. This requirement implies a significant stationarity condition in a non-cooperative environment. Along these lines, the strategic bargaining solution for long-run cooperation will be determined. Another interpretation will be given, too, which can interpret the strategic bargaining solution in terms of concessions. This section concludes the discussion of bargaining theory by summarizing some recent developments of strategic bargaining theory.

Throughout the main part of this section, the assumption will be made that both dU_i and dU_j are well defined along the set of efficient agreements. This assumption implies that the marginal rate of substitution, that is dU_i/dU_j, is well defined along the set of efficient agreements. It will be abandoned later because self-enforcing contracts may specify constraints (see (2.17) and (2.19)) which violate complete differentiability along the set of efficient agreements. It is, however, reasonable to start with this assumption and to discuss modifications later on.

The Nash Bargaining Solution

Until the seminal work of John Nash (1950, 1953), economic theory faced an indeterminacy problem for bilateral monopolies. The division of bargaining gains was completely indeterminate because the assumptions made for an agent's behaviour in a competitive environment were completely unsatisfactory and unable to resolve this ambiguity. Nash's approach was the first one to deal with this indeterminacy problem. His bargaining theory does not contain any detailed description of the bargaining process. Instead, he formulated several axioms which the bargaining solution should fulfil. It should be noted that Nash's approach does not consider utility functions as in Section 2.4 which has assumed that the agents' utilities are determined by the agreement. Instead, he merely assumed preference orderings on the set of lotteries over possible agreements. These orderings are assumed to satisfy the axioms of the von Neumann–Morgenstern expected utility theory. Thereby, Nash assumed that a function (a utility function in the sense of von Neumann and Morgenstern) can be defined for every agent which indicates preference for one lottery compared to another lottery if the expected utility of the first lottery is higher than the expected utility of the second one. The set of possible agreements, these utility functions and the disagreement point define the Nash bargaining problem.

This section will not go through all details of Nash's bargaining theory but will present Nash's axioms and the unique bargaining solution they imply. It will mainly follow Osborne and Rubinstein (1990) whose first three chapters are recommended to readers interested in a more thorough discussion of bargaining theory. Nash imposed four axioms which a bargaining solution should satisfy:

1. Invariance to equivalent utility representation: This axiom requires that a monotone transformation of the utility function (which gives the same preference ordering) should not alter the bargaining solution.
2. Symmetry: Nash did not consider different bargaining abilities because he required that all asymmetries should be taken into account by defining the

set of agreements and the disagreement point. Symmetry implies interchangeable agents and thereby identical bargaining gains for both agents.

3. Independence of irrelevant alternatives: This axiom states that every subset of the set of agreements which contains the bargaining solution of the larger set should give the same bargaining solution.
4. Pareto efficiency: This axiom requires that the solution belongs to the set of efficient agreements.

Let the von Neumann–Morgenstern utility functions be denoted by W_i and W_j. Nash demonstrated that the bargaining solution which satisfies all four axioms is given by

$$\prod_{k=i,j}\left[W_k\left(s_i^{Nash},s_j^{Nash}\right)-\Pi_k'\right] = \max_{s_i,s_j}\left\{\prod_{k=i,j}\left[W_k\left(s_i,s_j\right)-\Pi_k'\right]\right\}. \quad (2.33)$$

The Nash bargaining solution is unique and maximizes the product of both agents' bargaining gains. As Osborne and Rubinstein (1990) point out, the last two axioms carry some implicit assumptions concerning bargaining procedures. Nash himself discussed a demand game to introduce into the strategic aspects of bargaining. The strategic bargaining theory, however, experienced its breakthrough with the seminal paper of Rubinstein (1982). As strategic aspects are at the heart of this book, developing the strategic bargaining approach will be given more space (for overview articles reviewing non-cooperative bargaining theory, see Bester, 1989a; Binmore, Osborne and Rubinstein, 1992; Binmore, Rubinstein and Wolinsky, 1986; Sutton, 1986).

Strategic Bargaining

Every strategic bargaining approach assumes a certain structure of a bargaining game which specifies offers, acceptance and rejection in a bargaining process. This section will discuss the simplest variant of strategic bargaining, that is, the alternating offer game (see Table 2.2), but it will also demonstrate that the bargaining outcome can be predicted by this sequential model if the time which elapses between offers is negligibly small.

In Table 2.2, 0 indicates the earliest possible realization of bargaining gains. This alternating offer game has a finite horizon and assumes that agent i is to move first by making a proposal which is either accepted or rejected by agent j. If agent j accepts the proposal, the game is ended because an agreement was found. If he/she rejects i's proposal, both have to wait for one period after which it is up to agent j to make a proposal, and so on. In the last period, one agent makes a proposal and the other agent either accepts or rejects it, but is not able to make a counterproposal one period later. If acceptance of either agent never occurs, no agreement was reached.

Table 2.2 The alternating offer game

Time	Offer/ Reply
0	i/j
1	j/i
2	i/j
...	
TD	i/j if TD is even j/i if TD is odd
TD: terminal date	k/l reads: k makes an offer and l either accepts or rejects

Assume that the problem under consideration is simple because both agents have to agree on how to divide a pie of size one. This pie is available from 0 on and no other realization which concerns both agents is subject to bargaining. Without unanimous agreement, the share of the pie that both agents receive is zero. Let these shares be denoted by x_i and x_j, respectively. Assume further that the utilities both agents enjoy depend linearly on the share they receive:

$$U_i = x_i, U_j = x_j. \tag{2.34}$$

If a certain agreement is reached, the corresponding shares will be consumed immediately. Thus, this basic model of strategic bargaining assumes complete enforceability because neither agent is able to take a larger piece of the pie unless it was unanimously agreed upon. Let x^t =

$[x_i^t, x_j^t]$ denote a proposal in period t. The alternating bargaining game specifies that agent i makes a proposal, say x^0, in period 0. If it is accepted by agent j, the pie will be split accordingly. If it is rejected by agent j, it is up to him to make a proposal in period 1, say x^1. Obviously, both agents have opposite interests with respect to sharing the pie as i's utility is maximized by $[1,0]$ and j's utility is maximized by $[0,1]$.

What is the bargaining equilibrium in such an alternating offer game? Considering Nash equilibria, the result is ambiguous because every partition of the pie can stem from a Nash equilibrium. Suppose that agents i's strategy is always to offer $\hat{x}$ if it is his/her turn and to accept a proposal of j if and only if his/her proposed share does not fall short of $\hat{x}_i$. Agent j has a similar strategy to propose $\hat{x}$ and to reject every offer which does not give him/her at least $1 - \hat{x}_i$. In this case, $\hat{x}$ is a Nash equilibrium as no agent can improve on this outcome which will be realized in the first period. As $\hat{x}_i$ is not restricted, every agreement which is possible can be a Nash equilibrium and can be agreed upon in the first period. Other strategies may result in permanent disagreement.

However, these Nash equilibria involve a threat that is not credible. The seminal papers of Selten (1965, 1975) deal with the credibility of threats by discussing the behaviour at all nodes which are reachable in a game. His refinement of subgame perfection requires that only those equilibria are credible which are also equilibria for all subgames which are defined by the nodes reachable in the game. Consider, for example, period $TD - 1$, that is, the last but one bargaining period, and assume that $TD = 2$. In $TD - 1$, both agents know what will happen in the last period: in TD, agent i has all the bargaining power because he/she makes the final proposal and j can only reject or accept. Therefore, j accepts any offer of i because rejection ensures that both agents gain nothing. Assume that j votes for acceptance in the case of a zero share although in this case there is no difference between acceptance and rejection. Then, the Nash equilibrium in TD is given by $[1,0]$ because j is only able to reject or to accept but is not able to react by making a counterproposal. In period $TD - 1$, agent j knows that $[1,0]$ is the solution if no agreement is reached now and he/she knows that agent i is indifferent between δ_i now and 1 in the next period. Therefore, the proposal in $TD - 1$ is $[\delta_i, 1 - \delta_i]$. The same line of reasoning makes agent i propose $[1 - \delta_j(1 - \delta_i), \delta_j(1 - \delta_i)]$ in the first period which gives the subgame-perfect solution for the alternating-offer game with $TD = 2$:

$$x_i^*(TD = 2) = 1 - \delta_j(1 - \delta_i). \tag{2.35}$$

The star denotes the subgame-perfect outcome. It is obviously profitable to take the position of the last mover. However, the last-mover advantage disappears the longer the bargaining game lasts. If TD is even, the subgame-perfect equilibrium is given by

$$\forall \frac{TD}{2} \in \mathbb{N} : x_i^*(TD) = 1 - \sum_{t=1}^{TD} \delta_i^{t-1}\delta_j^t + \sum_{t=1}^{TD} \delta_i^t\delta_j^t . \qquad (2.36)$$

Now suppose that $TD = 1$. In this case, agent j is the last mover and

$$x_i^*(TD = 1) = 1 - \delta_j \qquad (2.37)$$

gives the subgame-perfect outcome because agent j is indifferent about receiving δ_j today and the whole pie tomorrow. If TD is odd, the subgame-perfect equilibrium is given by

$$\forall \frac{TD+1}{2} \in \mathbb{N} : x_i^*(TD) = \sum_{t=1}^{TD} \delta_i^{t-1}\delta_j^{t-1} + \sum_{t=1}^{TD} \delta_i^{t-1}\delta_j^t . \qquad (2.38)$$

In every case, the agreement is reached in period 0, the date for earliest realization, because both agents lose by waiting, and the pie is divided completely between both agents.

In Section 2.3, it was strongly argued that intertemporal problems should be tackled by assuming an infinite horizon. Obviously, an infinite horizon eliminates last-mover advantages. Equations (2.36) and (2.38) coincide for the limit case (for the first paper which developed this proof, see Binmore, 1987):

$$\lim_{TD \to \infty} x_i^*(TD) = \frac{1 - \delta_j}{1 - \delta_i\delta_j} . \qquad (2.39)$$

Equation (2.39) gives the solution for a typical Rubinstein model of alternating offers. Note that although the last-mover advantage is eliminated, there is still a first-mover advantage for agent i. If $\delta_i = \delta_j$ ($= \delta$) holds, x_i^* is $1/(1 + \delta)$ which exceeds 0.5 although both agents have identical discount factors. The first-mover advantage disappears the shorter the bargaining periods are.

Equation (2.39) gives the unique solution for strategic bargaining for one realization. It should be clear that the strategic bargaining solution is not changed when bargaining starts earlier than in 0 but the bargaining gains to be divided are not available earlier than in 0. The threat of delay is only credible when delay is possible, and delay is not possible before the pie is available. Thus, equation (2.39) also gives the bargaining solution for all cases in which communication between the two agents starts earlier because they anticipate the bargaining procedure and the threat of delays when the pie is available. This feature is also the basis for bargaining for long-run cooperation because bargaining before realizations can occur does not change the bargaining result.

Bargaining for Long-run Cooperation

Strategic bargaining for long-run cooperation makes sense only if delay of every realization is possible. If delay were not possible, any agent could successfully offer a proposal just before the realization occurs which gives the maximum bargaining gains subject to the compliance of his/her partner. Hence, (2.15) is to be replaced by a more general assumption. Every action taken by an agent is restricted to a certain time structure so that actions which refer to period σ, $\sigma \in \mathbb{N}_0$, cannot be realized before σ units of time have elapsed from the start of the game. The minimum period between two realizations is one:

$$\forall \sigma \in \mathbb{N}_0 : r(\sigma) \geq \sigma, \ r(\sigma+1) - r(\sigma) \geq 1. \tag{2.40}$$

$r(\sigma)$ denotes σ's realization, that is, the realization which is not possible before σ units of time have elapsed and before at least one unit has elapsed after realization $r(\sigma - 1)$. Assumption (2.40) is not as trivial as it seems at first glance because it defines solely a lower limit for realizations in an intertemporal setting. It should be noted that certain actions taken for period σ can at best be realized after σ periods have elapsed, but it is also possible that realization takes place later, thereby implying a delay for all following periods. Assumption (2.40) is more general than the assumptions made in the repeated games literature, and it allows for the threat of delay in strategic bargaining. The assumption that the earliest realizations are integers calibrates the time which must elapse between a preceding and a following earliest realization.

When two agents bargain for long-term cooperation the time structure of which meets (2.40), they not only consider the payoffs possible in the next period but the whole stream of payoffs. Due to (2.40), delay in the first period causes delay in all following periods as well. As the threat of delays will be anticipated by both agents, they may bargain either for some realizations or for the whole stream of realizations. The first case is discussed by Muthoo (1995) who assumes that only one (enforceable) realization is subject to current bargaining, and the next realization is subject to bargaining after previous realization has occurred. However, it is not explained why two agents restrict bargaining to one realization when all future options are known. Additionally, separate bargaining is not an appropriate assumption in a non-cooperative environment (for a discussion of Muthoo's paper, see Stähler, 1996b). Because the profitability of self-enforcing contracts compared to defection relies on future results, every agent must know what future results from current agreements look like. This requirement does not imply that factual bargaining covers all future realizations but that bargaining results are derived as if bargaining covered all future realizations.

The general setting of this chapter implies that bargaining and potential delay are not restricted to bargaining gains. Consider two firms in a duopolistic market which bargain for implicit collusion. They are not able to realize the non-cooperative profits as soon as possible and then some bargaining gains later on. Realizing the non-cooperative payoffs and realizing bargaining gains are not separable. Therefore, any delay in bargaining affects the total profits and not just the bargaining gains (for a general discussion, see Stähler, 1997). When two firms bargain for long-run cooperation, their total profits are subject to the threat of delay. Hence, any firm will enter into a bargaining process only if the expected bargaining gains fall short of delay costs. As it will be shown that delays will never occur, and since bargaining gains must be strictly positive in order to guarantee compliance, every firm can be assumed to enter into a bargaining process.

When proposals cover explicitly or implicitly all future realizations, they will specify the earliest realizations which are possible at the very moment of the proposal. This is the counterpart of the one-shot bargaining game in which no proposal is expected to specify any further delay beyond the very moment of the proposal. Let $\hat{\pi}_k(m)$ indicate the respective prospect of strategic bargaining for a long-run cooperation when long-run cooperation starts in m. $\hat{\pi}_k(m)$ denotes the sum of discounted present and future payoffs, and $s_k(t)$ denotes the strategy level realized in t:

$$\forall k \in \{i, j\}, \forall m \in \mathbb{N}_0 : \hat{\pi}_k(m) := \sum_{t=m}^{\infty} \delta_k^{[r(m)-m+t]} \Pi_k \big[s_i(t), s_j(t) \big]. \quad (2.41)$$

$\hat{\pi}_k(m)$ gives the utility sum which is discounted on the earliest start of realization m and not on the first factual realization $r(m)$ because the mth realization is not necessarily realized after m units of time have elapsed. If delay had occurred before $r(m) = m$, realizations had to take place later. For example, consider two firms bargaining for implicit collusion, which is repeated every year. If a delay of, say, one week had occurred, all other realizations are delayed for at least one week as well so that $r(m) > m$. Only if delay never occurs, $\hat{\pi}_k[m]$ is defined by

$$\sum_{t=m}^{\infty} \delta_k^t \Pi_k \left[s_i(t), s_j(t) \right].$$

When both agents enter into a bargaining process for long-run cooperation, they bargain for the whole future stream of potential realizations and may specify different strategy profiles in the course of time if agreements were enforceable. In a non-cooperative environment, however, the lack of any commitment implies a significant stationarity result.

Consider the case that bargaining for long-run cooperation would imply different policies for different periods: suppose that a contract specifies different strategy levels for different periods 1 and 2,

$$\left[s_i(1), s_j(1), s_i(2), s_j(2) \right], \quad s_i(1) \neq s_i(2), \quad s_j(1) \neq s_j(2).$$

Without loss of generality, assume that agent i (j) receives higher (lower) payoffs in the first period compared to the second period:

$$\Pi_i \left[s_i(1), s_j(1) \right] > \Pi_i \left[s_i(2), s_j(2) \right], \quad \Pi_j \left[s_i(1), s_j(1) \right] < \Pi_j \left[s_i(2), s_j(2) \right]$$

This case is shown in Figure 2.3.

After the first period's policy has been realized, however, the situation is the same as before the first period. If the situation is the same as before, why should agent i now keep his/her promises which were made in the previous bargaining procedure? Instead, if both agents accept that bygones are bygones and both recognize that no agent has broken the contract, there is no reason why communicating about a contract starting in period 2 after the contract has been realized according to $[s_i(1), s_j(1)]$ should give different results from bargaining for a self-enforcing contract starting in period 1.

When every agent can submit proposals which may successfully replace existing agreements, bargaining behaviour in a non-cooperative environment has to be time consistent. A bargaining-based agreement is time consistent if it is not subject to successful revision. In such a non-cooperative setting, bargaining for identical streams of potential realizations in time should lead to time-independent bargaining results because no agent can credibly commit him-/herself not to exploit his/her bargaining power in the future. As the situation is the same as before the first realization (except when one agent has defected), the agreed-upon strategies are identical in time. This stationarity result implies that the discounted utility can be given in terms of s_i and s_j only. Hence, time-consistent bargaining behaviour implies that the strategy levels must not change in the course of time. The lack of credible commitments does not allow non-stationary strategies to be specified.

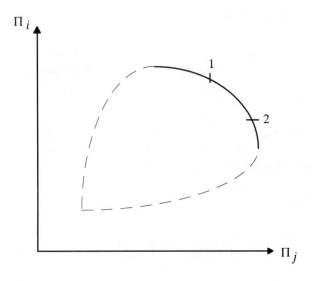

Figure 2.3   Different payoff combinations in different periods

The condition of time-consistent bargaining behaviour resembles Markov perfection. It requires that the bargaining results should depend only on a state variable which signals whether an agent has defected in the past (for a discussion of Markov perfection and cooperation in repeated games, see Stähler 1996c). If no agent has defected or – for whatever reason – behaviour has not been coordinated by a self-enforcing contract, bargaining should lead to time-independent results. In this sense, time-consistent bargaining behaviour requires that bargaining should not depend on the history of previous bargaining.

When the strategy levels do not change in the course of time, strategic bargaining for a self-enforcing contract translates into bargaining for a single strategy level for all future realizations. Let h denote the delay from realizing s_i and s_j in h instead of 0:

$$\hat{\Pi}_k(s_i,s_j):=\frac{1}{1-\delta_k}\Pi_k(s_i,s_j)\Rightarrow\hat{\pi}_k(0)=\delta_k^h\hat{\Pi}_k(s_i,s_j). \qquad (2.42)$$

$\hat{\Pi}_k(s_i, s_j)$ defines the discounted sum of current and future payoffs without delay of realizations. Because of the requirement of time-consistent bargaining behaviour, $\hat{\Pi}_k$ may replace $\hat{\pi}_k$ when the strategic bargaining solution is to be determined.

Strategic Bargaining for Long-run Cooperation

Developing the strategic bargaining solution assumes that the alternating offer game which defines strategic bargaining is still given by Table 2.2 but the time which elapses between offer and counter offer (that is, the bargaining period) will be denoted by h and $TD \rightarrow \infty$. Both agents bargain for the future stream of realizations before the first realization can occur, but they anticipate the potential threats of delay. When they have unanimously agreed upon a certain policy, they are supposed to move simultaneously and directly when realization is possible. It is still assumed that agent i has the first-mover advantage because it is up to him/her to propose when the dates of earliest realization are reached. Then, agent i has to make a proposal in the bargaining periods

$$0, 2h, 4h, \ldots, th, \ldots, \quad \frac{t}{2}\in\mathbb{N}_0$$

and has to decide on accepting or rejecting agent j's proposals in the bargaining periods

$$h, 3h, \ldots, [t+1]h, \ldots, \quad \frac{t}{2}\in\mathbb{N}_0.$$

This generalized model will argue along the lines of Shaked and Sutton (1984). This paper demonstrated that the unique subgame-perfect bargaining solution can be developed by determining the maximum and the minimum bargaining gains of an agent. Shaked and Sutton assume (as do other authors) that payoffs are zero when no bargaining takes place. As the non-cooperative payoffs are not necessarily zero and delay concerns total payoffs, strategic bargaining for payoffs and not only for bargaining gains must be considered.

Accordingly, let $[s_i^*, s_j^*]$ denote the strategy levels which ensure a certain payoff agent i can receive in a subgame-perfect equilibrium, for example, the maximum or minimum payoff. Consider bargaining period th which involves a delay of th from the respective earliest realization, and assume that agent i receives the respective payoff $\hat{\Pi}_i[s_i^*, s_j^*]$. $\hat{\Pi}_i[s_i^*, s_j^*]$ gives the discounted sum of all maximum payoffs in the future when they are discounted on th. In the preceding bargaining period $(t-1)h$, agent j makes a proposal and agent i accepts every proposal which gives him/her at least $\delta_i^h \hat{\Pi}_i[s_i^*, s_j^*]$. Let $[s_i^{**}, s_j^{**}]$ denote the subgame-perfect strategy levels in period $(t-1)h$ which are given by

$$\delta_i^h \hat{\Pi}_i\left[s_i^*, s_j^*\right] = \hat{\Pi}_i\left[s_i^{**}, s_j^{**}\right],$$

$$\hat{\Pi}_j\left[s_i^{**}, s_j^{**}\right] = \max\left\{\hat{\Pi}_j\left[s_i, s_j\right] \text{ s.t. } \delta_i^h \hat{\Pi}_i\left[s_i^*, s_j^*\right] = \hat{\Pi}_i\left[s_i^{**}, s_j^{**}\right]\right\}.$$

$$(2.43)$$

Equation (2.43) contains two dependent conditions. First, agent i is made indifferent between realization today or tomorrow. Second, agent j will propose a strategy combination which maximizes his/her payoffs subject to the first condition (note that several $s_i - s_j$ combinations exist to which agent i is indifferent).

In period $(t-2)h$, agent j will not accept any proposal which does not give him/her at least the payoff $\hat{\Pi}_j[s_i^{**}, s_j^{**}]$, modified by the concessions which are due to the preference for an earlier realization. Thus, agent i obtains the maximum payoff by proposing $[s_i^{***}, s_j^{***}]$ which equalizes the discounted minimum payoff agent j would be given in $(t-1)h$:

$$\delta_j^h \hat{\Pi}_j\left[s_i^{**}, s_j^{**}\right] = \hat{\Pi}_j\left[s_i^{***}, s_j^{***}\right],$$

$$\hat{\Pi}_i\left[s_i^{***}, s_j^{***}\right] = \max\left\{\hat{\Pi}_i\left[s_i, s_j\right] \text{ s.t. } \delta_j^h \hat{\Pi}_j\left[s_i^{**}, s_j^{**}\right] = \hat{\Pi}_j\left[s_i^{***}, s_j^{***}\right]\right\}$$

$$(2.44)$$

This bargaining procedure and its results are shown in Table 2.3. As it is assumed that agent i is always the first mover, $(t-2)h$ may be set zero and is equal to the dating of the earliest realizations.

Table 2.3 assumes that a certain payoff combination of the set of efficient agreements can be implied only by a unique strategy pair. The game started in th is the same game started in $(t-2)h$, and subgame perfection requires that strategies should constitute a Nash equilibrium in all subgames they can reach. If the game in period $(t-2)h$ is the same as in period th, the corresponding solutions must be identical, and $[s_i^{***}, s_j^{***}]$ must be equal to $[s_i^*, s_j^*]$ if a certain payoff combination of the set of efficient agreements is implied by one and only one strategy pair.

Table 2.3 Payoffs in different bargaining periods

Bargaining periods	Strategic bargaining variables	Payoff of agent i	Payoff of agent j
$(t-2)h$	$\left[s_i^{***}, s_j^{***}\right]$	$\hat{\Pi}_i\left[s_i^{***}, s_j^{***}\right]$ maximizes agent i's payoff s.t. indifference of agent j	$\delta_j^h \hat{\Pi}_j\left[s_i^{**}, s_j^{**}\right]$ $= \hat{\Pi}_j\left[s_i^{***}, s_j^{***}\right]$
$(t-1)h$	$\left[s_i^{**}, s_j^{**}\right]$	$\delta_i^h \hat{\Pi}_i\left[s_i^*, s_j^*\right]$ $= \hat{\Pi}_i\left[s_i^{**}, s_j^{**}\right]$	$\hat{\Pi}_j\left[s_i^{**}, s_j^{**}\right]$ maximizes agent j's payoff s.t. indifference of agent i
th	$\left[s_i^*, s_j^*\right]$	$\hat{\Pi}_i\left[s_i^*, s_j^*\right]$	$\hat{\Pi}_j\left[s_i^*, s_j^*\right]$
subgame-perfection	$\left[s_i^*, s_j^*\right] = \left[s_i^{***}, s_j^{***}\right]$		

$[s_i^*, s_j^*]$ must belong to the set of efficient agreements. If it did not, it could not define a perfect equilibrium because one agent would accept an alternative proposal which gives at least the same payoff and the other agent could be better off. If $[s_i^*, s_j^*]$ belongs to the set of efficient agreements, $[s_i^{**}, s_j^{**}]$ also belongs to this set because it maximizes agent j's payoffs subject to making agent i indifferent about realization today or tomorrow.

Until now, it has not been shown under which conditions only a unique $[s_i^*, s_j^*]$ satisfies (2.43) and (2.44). These conditions will be demonstrated below when the bargaining solution is interpreted in terms of concessions.

Interpreting the Bargaining Solution in Terms of Concessions

Conditions (2.43) and (2.44) cannot be solved explicitly without simple functional specification. But focusing on small bargaining periods allows the strategic bargaining problem in terms of concessions to be reformulated. The idea is that agent i is prepared to accept certain concessions, that is, a certain change of the strategy combination, if he/she can realize agreements in $(t - 1)h$ when agent j makes a proposal. Similarly, agent j is prepared to accept certain concessions, that is, a certain change in the opposite direction, if he/she can realize agreements in $(t - 2)h$ when agent i makes a proposal. Subgame perfectness requires that neither concession changes either agent's payoffs so that no agent can gain by waiting. In Table 2.3, agent i's concession was to allow the move from $[s_i^*, s_j^*]$ to $[s_i^{**}, s_j^{**}]$ and agent j's concession was to allow the move from $[s_i^{**}, s_j^{**}]$ to $[s_i^{***}, s_j^{***}]$ which is equal to $[s_i^*, s_j^*]$ in a subgame-perfect bargaining equilibrium. Thus, the payoff changes from th to $(t - 2)h$, which result from the equilibrium concessions, just compensate each other. If these changes implied by the concessions fell apart, one agent could gain by waiting for his/her next offer opportunity.

Let $\Delta\hat{\Pi}_i$ and $\Delta\hat{\Pi}_j$ denote the equilibrium concessions which agent i and agent j accept for an earlier realization:

$$\hat{\Pi}_i\left[s_i^{**}, s_j^{**}\right] = \hat{\Pi}_i\left[s_i^*, s_j^*\right] + \Delta\hat{\Pi}_i,$$

$$\hat{\Pi}_j\left[s_i^{**}, s_j^{**}\right] = \hat{\Pi}_j\left[s_i^*, s_j^*\right] + \Delta\hat{\Pi}_j. \tag{2.45}$$

Conditions (2.45) imply

$$\delta_i^h\hat{\Pi}_i\left[s_i^*, s_j^*\right] = \hat{\Pi}_i\left[s_i^*, s_j^*\right] + \Delta\hat{\Pi}_i,$$

$$\delta_j^h\left[\hat{\Pi}_j\left[s_i^*, s_j^*\right] + \Delta\hat{\Pi}_j\right] = \hat{\Pi}_j\left[s_i^*, s_j^*\right]. \tag{2.46}$$

$\Delta \hat{\Pi}_i$ is negative, $\Delta \hat{\Pi}_j$ is positive. Since both $[s_i^*, s_j^*]$ and $[s_i^{**}, s_j^{**}]$ belong to the set of efficient agreements, $\Delta \hat{\Pi}_i / \Delta \hat{\Pi}_j$ is well defined along the set of efficient agreements. Division of the second line of (2.46) by its first line yields

$$\frac{\Delta \hat{\Pi}_i / \hat{\Pi}_i \left[s_i^*, s_j^* \right]}{\Delta \hat{\Pi}_j / \hat{\Pi}_j \left[s_i^*, s_j^* \right]} = \frac{\delta_j^h \left(\delta_i^h - 1 \right)}{1 - \delta_j^h}. \tag{2.47}$$

From (2.46), it can be seen that the equilibrium concessions are smaller in absolute terms the shorter the bargaining period is. Hence, $\Delta \hat{\Pi}_i / \Delta \hat{\Pi}_j$ may be approximated by $d \hat{\Pi}_i [s_i^*, s_j^*] / d \hat{\Pi}_j [s_i^*, s_j^*]$ for small bargaining periods which imply sufficiently small equilibrium concessions:

$$\frac{d \hat{\Pi}_i / \hat{\Pi}_i \left[s_i^*, s_j^* \right]}{d \hat{\Pi}_j / \hat{\Pi}_j \left[s_i^*, s_j^* \right]} = \frac{\delta_j^h \left(\delta_i^h - 1 \right)}{1 - \delta_j^h}. \tag{2.48}$$

According to (2.42),

$$\forall k \in \{i, j\} : d \hat{\Pi}_k \left[s_i, s_j \right] = \frac{1}{1 - \delta_k} d \Pi_k \left[s_i, s_j \right] = \frac{1}{1 - \delta_k} d U_k \left[s_i, s_j \right] \tag{2.49}$$

holds. Consequently, the discount factor terms drop out of (2.48) and give

$$\frac{d U_i / \Pi_i \left[s_i^*, s_j^* \right]}{d U_j / \Pi_j \left[s_i^*, s_j^* \right]} = \frac{\delta_j^h \left(\delta_i^h - 1 \right)}{1 - \delta_j^h}. \tag{2.50}$$

Thus, strategic bargaining for long-run cooperation in a non-cooperative environment yields the same strategy levels for each period as strategic bargaining for one realization if contracts were enforceable, given that the set of efficient agreements is the same in both cases. This is a salient result which is due to the stationarity of long-run cooperation implied by time-consistent bargaining behaviour. Note carefully that the identity of results assumes that both sets of efficient agreements are identical. This result does not hold when the sets of efficient agreements differ.

In the standard bargaining games of splitting a pie of size one, bargaining gains are linear with respect to the shares an agent receives, that is, $dU_i/dU_j = -1$, and $U_j = \Pi_i = 1 - U_i = 1 - \Pi_j$ due to the unchanged size of the pie in terms of total bargaining gains. Additionally, the non-cooperative equilibrium, that is, the result in the absence of any bargaining, implies zero payoffs. Inserting these assumptions into (2.50) (and observing that linearization gives exact results in that case) produces the well-known formula of Rubinstein's paper and reproduces (2.39) for $h = 1$:

$$-\frac{1-U_i}{U_i} = \frac{\delta_j^h\left(\delta_i^h - 1\right)}{1-\delta_i^h} \Leftrightarrow U_i = \frac{1-\delta_j^h}{1-\delta_i^h\delta_j^h}. \tag{2.51}$$

As the approach adopted in this book focuses on the role of discount factors and not on mover advantages, it does not merely assume a sufficiently small h to allow the approximation of $\Delta\hat{\Pi}_k$ by $d\hat{\Pi}_k$ but also that bargaining periods are so short that $h \to 0$ may be assumed. This assumption should not indicate that bargaining is not based on offers and counter offers but it indicates that responses to offers are extremely quick and can be approximated by $h \to 0$. Applying L'Hôpital's Rule to the RHS of (2.50) gives the general formula of a subgame-perfect bargaining equilibrium for infinitely small bargaining periods:

$$\rho\left[s_i^*, s_j^*\right] := \frac{dU_i/\Pi_i\left[s_i^*, s_j^*\right]}{dU_j/\Pi_j\left[s_i^*, s_j^*\right]} = -\frac{\ln\delta_i}{\ln\delta_j}. \tag{2.52}$$

Condition (2.52) shows that the relative change of agent i's payoffs divided by the relative change of agent j's payoffs must be equal to a ratio of waiting costs in a perfect bargaining equilibrium. The first term will be denoted by ρ in the remainder of the book and gives the elasticity of payoffs which describes the percentage change of agent i's payoffs if agent j's payoffs are changed by one percentage. Condition (2.52) has an obvious economic interpretation: in a subgame-perfect equilibrium, the more impatient (patient) agent is already (only) happy with a result which is caused by small (great) relative payoff changes of the other agent which increase his/her relative payoff because he/she has to balance a further denial of proposals with the very costly (less costly) delay. ρ mirrors the technical aspects of the bargaining problem and $\ln\delta_i/\ln\delta_j$ mirrors the relative impatience. Balancing both terms follows from the condition that the bargaining gains of each agent should not be changed by marginal concessions of both agents in a subgame-perfect bargaining equilibrium. When the model uses the simple linear bargaining gain functions which standard models employ, that is, $U_i = x_i$, $U_j = 1 - x_i$, the LHS of (2.52) is $-(1 - x_i)/x_i$ and the subgame-perfect equilibrium merely balances the marginal impatience of both agents. A similar model can be found in Chae (1993).

One may now turn to cases in which the strategic bargaining solution is unique. Consider Figure 2.4, which gives a concave and differentiable frontier of efficient agreements.

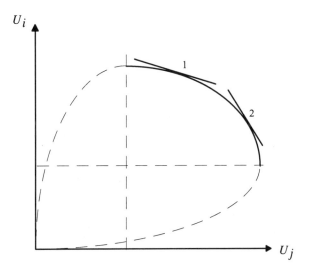

Figure 2.4   Frontier of the set of efficient agreements I

Concavity must not be strict but the frontier may be linear as well. From Figure 2.4, an important implication of a concave frontier of the set of efficient agreements can be derived. Consider the two points 1 and 2 on the Pareto frontier. It is evident that Π_i/Π_j is increased when moving from 2 to 1. Additionally, the depicted slopes at 1 and 2 show that the marginal rate of substitution $d\Pi_i/d\Pi_j$ is increased when moving from 2 to 1 (or remains unchanged if the frontier is linear). Hence, one finds that the elasticity

$$\rho = \frac{d\Pi_i/d\Pi_j}{\Pi_i/\Pi_j} = \frac{d\Pi_i/\Pi_i}{d\Pi_j/\Pi_j}$$

increases when moving from 2 to 1 (note that $d\Pi_i/d\Pi_j$ is negative). Since Figure 2.4 assumes that the frontier of a set of efficient agreements is continuously differentiable, ρ decreases without any jump from the furthest point right to the furthest point left along the frontier of the set of efficient agreements.

In this case, the relationship between the payoff levels and the discount factor combination is unique: one and only one payoff combination is implied by a certain discount factor combination and one and only one discount factor combination implies a certain payoff combination: since $-\ln\delta_i/\ln\delta_j$ lies in the range $]-\infty, 0]$, uniqueness of the strategic bargaining solution is guaranteed: when moving leftwards along the set of efficient agreements, Π_i/Π_j is increased and $d\Pi_i/d\Pi_j$ is increased. On the furthest point right of the set of efficient agreements point $d\Pi_i/d\Pi_j$ approaches $-\infty$, on the furthest point left $d\Pi_i/d\Pi_j$ is zero. Hence, ρ is increased continuously from $-\infty$ to 0 when moving leftwards along the set of efficient agreements. No strategy combination exists in the set of efficient agreements which implies the same elasticity. If this payoff combination is implied by one pair of strategies only, uniqueness in strategies is guaranteed as well.

Figure 2.4 shows a bargaining problem for which the relevant bargaining solutions do not cover the whole range of positive bargaining gains. This result is due to compliance constraints. In the case of enforceable contracts, the payoff frontier can be expected to be concave in the whole range of positive bargaining gains. On the frontier, smoothness is guaranteed because every point on the frontier has a unique tangent $dU_i(.)/dU_j(.)$ like the one drawn in point 1 and point 2.

Dropping smoothness allows multiple discount factor combinations to imply the same bargaining outcome. This result will be shown to have strong relevance in the next chapters as well. Figure 2.5 illustrates this case. In Figure 2.5, point J indicates a non-differentiable point because $dU_i(.)/dU_j(.)$ does not exist at J but jumps when moving along the frontier of efficient agreements. If J is held fixed, the payoffs Π_i and Π_j are fixed, too, and dU_i/dU_j is determined by the discount factors:

$$\frac{dU_i}{dU_j} = -\frac{\ln\delta_i}{\ln\delta_j}\frac{\Pi_i}{\Pi_j}$$

As the jump implies that all tangents of the slope which lies in the range between those drawn in Figure 2.5 imply point J, there is more than one discount factor combination which implies J as the strategic bargaining solution. In this case, uniqueness of the bargaining solution is also guaranteed because no strategy combination exists which implies the same elasticity, but the same elasticity is not increased continuously when moving leftwards along the set of efficient agreements. In the remainder of this book, complete differentiability will not be assumed, so as to allow for the discussion of bargaining problems as depicted by Figure 2.5.

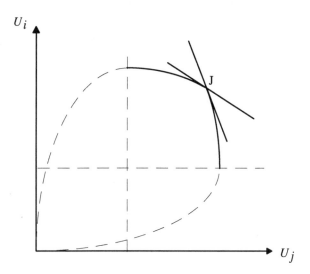

Figure 2.5   Frontier of the set of efficient agreements II

Strategic bargaining theory has received increasing attention since Rubinstein's paper, and it should be noted that the strategic bargaining model leading to (2.52) relies especially on the assumption of perfect information about the payoffs and the discount factor. Other models explore the behaviour of bargaining agents under uncertainty which may result in delay. Many papers deal with different aspects of uncertainty and incomplete information (Admati and Perry, 1987; Ausubel and Deneckere, 1992; Bikhchandani, 1992; Cramton, 1992; Gul and Sonnenschein, 1988; Rubinstein, 1985a,b). Delay is possible because an agent may learn the attributes of his/her bargaining partner in the course of time and has to balance expected gains from uncovering some private information about his/her opponent and an early realization (for an overview article on bargaining with private information, see Kennan and Wilson, 1993). Other papers consider transaction costs of bargaining (Cramton, 1991), randomize the offer–counter offer structure (Hoel, 1987), or discuss deadlines and imperfect control (Ma and Manove, 1993). All these approaches make more realistic assumptions with respect to information sets or the structure of the alternating offer game. However, if the bargaining periods shrink, the probability of delay shrinks as well because all relevant information is available earlier. If the bargaining period approaches zero, delays will never occur because all information can be obtained promtly.

As the present approach does not involve the impact of strict mover structures, these will not be discussed here. Additionally, assuming information asymmetries would also set the stage for strategies which build up reputation. Reputation is based on certain subjective probabilities that a certain agent is of a certain type. These probabilities may be influenced by the actions of this agent in order to build up a certain reputation. Although reputation obviously plays a significant role in real-life cooperation, its analytic treatment suffers from the need to specify priors. The specification of priors, that is, subjective probabilities for the beginning of the game, has a strong influence on the result. In fact, every possible outcome can be produced by a certain prior specification. As one of the aims of this book is to resolve the ambiguity problem of possible cooperation in a non-cooperative environment, the assumption of almost complete information will not be relaxed.

2.6   ON INTERTEMPORAL EFFECTS OF STRATEGIC BARGAINING FOR A SELF-ENFORCING CONTRACT

After the model of strategic bargaining for self-enforcing contracts has been developed, the following chapters will apply the model to different applications. Section 2.4 has already demonstrated that compliance constraints depend on the impatience of both agents who have to balance the instantaneous gains of non-compliance, future punishments and the gains from long-run cooperation. The preceding section has demonstrated that the universal option to bargain at every moment does not allow non-stationary joint policies to be realized. This result limits the scope of agreements significantly as strategic bargaining for a self-enforcing contract implies stationary contracts.

This section concentrates on the result of restricting self-enforcing contracts to stationary contracts and discusses the corresponding welfare implications for two agents. It begins with a discussion on the division of bargaining gains of coordinated actions, the repetition of which is infinite if every contract is enforceable. Examples are given for two different payoff functions which demonstrate the intertemporal efficiency conditions for enforceable long-run contracts, that is, those which cover all future periods. Then, these contracts are compared with an infinite repetition of stationary contracts. The analysis will focus on the intertemporal effects and neglect compliance constraints, which are dealt with explicitly in the following chapters.

The discount factors not only determine the division of bargaining gains but they also signal the intertemporal preference of both agents. When utility is transferable and neither agent faces transfer restrictions which depend on the size of utility to be transferred (for example, credit market restrictions), they consider the sum of all individual bargaining gains which are denoted by BG and given by

$$BG = \sum_{t=0}^{\infty} \left\{ \delta_i^t U_i \left[x_i(t) \right] + \delta_j^t U_j \left[x_j(t) \right] \right\}, \forall k \in \{i, j\} : \frac{dU_k}{dx_k(t)} > 0. \quad (2.53)$$

Equation (2.53) contains the assumption that bargaining gains can be transferred to the other agent and enter both agents' utility function by simple addition (the same assumption will be made in Chapter 4). As in Section 2.5, x_i and x_j denote the shares the agents receive, and it is assumed that the non-cooperative payoffs are zero. Not considering compliance constraints assumes without loss of generality that these shares add up to unity if the pie is completely divided. This assumption simplifies the development of two efficiency requirements which enforceable long-run contracts should fulfil in order to guarantee Pareto efficiency. Note that Pareto efficiency does not predetermine the division of bargaining gains but ensures that the total bargaining gains to be divided are maximized. In this case, both agents prefer the maximum of the total gains even if an agent's individual bargaining gain is small but his/her low share also increases with joint bargaining gains.

Pareto efficiency requires static efficiency which ensures that the whole pie is divided in each period. Any pie left over in one period implies forgone utility for both agents:

$$x_j(t) = 1 - x_i(t) \Rightarrow \forall t : \frac{dx_j(t)}{dx_i(t)} = -1. \quad (2.54)$$

The more salient feature of enforceable long-run contracts is intertemporal efficiency. Both agents discount the shares they receive in the future and both agents are better off if the more patient agent receives a higher share in later periods and the more impatient agent receives a higher share in earlier periods. Because the transfer of utility is not restricted, both agents maximize their joint bargaining gains (2.53). Maximization of (2.53) under the use of the static optimality condition (2.54) leads to

$$\forall t:\ \delta_i^t dU_i(t) - \delta_j^t dU_j(t) \le 0,$$

$$\left[\delta_i^t dU_i(t) - \delta_j^t dU_j(t) \right] \cdot x_i(t) \cdot \left[1 - x_i(t) \right] = 0. \tag{2.55}$$

Specific utility function can stress the importance of intertemporal efficiency for long-run contracts. Consider the net utility functions (2.34) which assumed linear bargaining gains. If the game is repeated infinitely, the refrigerator of both agents is endowed with a pie of size one in each period. Inserting (2.34) into the intertemporal efficiency condition (2.55) shows that the more patient agent should receive the whole pie in all periods and compensate the other agent:

$$\forall t,\ \forall k \in \{i, j\}:\ x_k(t) = 1,\ x_{-k}(t) = 0 \ \text{ for } \ \delta_k > \delta_{-k}.$$

This result is hardly surprising because the more impatient agent shows a superior intertemporal efficiency and should receive the whole pie in all periods. When $\delta_i = \delta_j$, the intertemporal division of the pie does not play any role because the total bargaining gains are already maximized by the static efficiency condition.

The logarithmic net utility functions

$$U_i\big[x_i(t)\big] = \ln\big[x_i(t)\big],\ \ \frac{dU_i}{dx_i(t)} = \frac{1}{x_i(t)},$$

$$U_j\big[x_j(t)\big] = \ln\big[x_j(t)\big],\ \ \frac{dU_j}{dx_j(t)} = \frac{1}{x_j(t)}$$

imply an intertemporal efficiency condition according to

$$\forall t:\ \frac{1 - x_i(t)}{x_i(t)} = \left(\frac{\delta_j}{\delta_i}\right)^t.$$

In this case, intertemporal efficiency implies an interior solution for all periods because decreasing marginal utilities rule out corner solutions. The more impatient (patient) agent receives a lower (higher) share in the course of time. If the solution paths do not imply corner solutions, rearranging (2.55) in order to isolate the discount factor of agent j gives

$$\delta^t_j = -\frac{dU_i\big[x_i(t)\big]}{dU_j\big[x_j(t)\big]}\,\delta^t_i$$

which can be used to determine the maximum total bargaining gains BG^{max}:

$$BG^{max} = \sum_{t=0}^{\infty}\delta^t_i U_i\big[x_i(t)\big]\left\{1 - \frac{dU_i\big[x_i(t)\big]}{dU_j\big[x_j(t)\big]}\frac{U_i(t)}{U_j(t)}\right\} = \sum_{t=0}^{\infty}\delta^t_i U_i\big[x_i(t)\big]\{1-\rho(t)\}$$

(2.56)

$\rho(t)$ denotes the bargaining gain elasticity of a long-term contract in t. Let U_i^* (U_j^*) denote the bargaining gains agent i (j) receives from an enforceable long-run contract according to the strategic bargaining solution:

$$U_i^* = \frac{\ln\delta_j}{\ln\delta_i + \ln\delta_j}\,BG^{max}, \quad U_j^* = \frac{\ln\delta_i}{\ln\delta_i + \ln\delta_j}\,BG^{max}.$$

Stationary contracts, however, imply a constant bargaining gain elasticity:

$$\forall t \geq 0: \ \frac{dU_i/U_i}{dU_j/U_j} = \text{const.} \Rightarrow BG(\rho = \text{const.}) \leq BG^{max}.$$

This restriction lowers both the total and the individual bargaining gains because it prevents both agents from realizing discount factor-driven intertemporal redistribution which raises the total bargaining gains. Only in the case of identical discount factors, that is, $\delta_i = \delta_j$, are both agents indifferent between long-term and stationary contracts. Generally, both agents are weakly worse off with stationary contracts because they separate the periods for which the agents are able to bargain, and prevent intertemporal utility maximization. This is obvious for the case of logarithmic utility functions when the increasing (decreasing) share for the more patient (impatient) agent is replaced by fixed shares for all periods.

The model defined by (2.34) makes only the impatient agent strictly worse off for switching from a long-term contract to a stationary one. Let U_i^s (U_j^s) denote the bargaining gains, that is, the discounted sum of gains, which agent i (j) receives from an infinitely repeated (and enforceable) stationary contract according to the strategic bargaining approach. In every period, the division of the pie is given by $\ln\delta_j/(\ln\delta_i + \ln\delta_j)$ and $\ln\delta_i/(\ln\delta_i + \ln\delta_j)$ for agent i and agent j, respectively. The discounted sums U_i^s and U_j^s are given by

$$U_i^s = \frac{\ln \delta_j}{\ln \delta_i + \ln \delta_j} \frac{1}{1-\delta_i},$$

$$U_j^s = \frac{\ln \delta_i}{\ln \delta_i + \ln \delta_j} \frac{1}{1-\delta_j}. \qquad (2.57)$$

Suppose that $\delta_i > \delta_j$ holds and induces $x_i(t) = 1$ for all periods when long-term contracts are possible. The maximum total bargaining gains are given by $1/(1 - \delta_i)$ due to agent i's relatively lower impatience. Hence, the bargaining gains for the long-term contract (denoted by the superscript l) are

$$U_i^l = \frac{\ln \delta_j}{\ln \delta_i + \ln \delta_j} \frac{1}{1-\delta_i},$$

$$U_j^l = \frac{\ln \delta_i}{\ln \delta_i + \ln \delta_j} \frac{1}{1-\delta_i}. \qquad (2.58)$$

Equation (2.58) shows that switching to stationary contracts does not decrease the discounted sum of agent i's bargaining gains but that of agent j. In general, however, switching from long-term to stationary contracts makes both agents worse off if utility functions are not linear. Hence, the conditions of time-consistent bargaining behaviour and a non-cooperative environment restrict the scope for mutual improvements in two ways: on the one hand, intertemporal reallocation is impossible, on the other hand, weak renegotiation-proofness imposes compliance constraints. The first point was explicitly discussed in this section since it is likely to be overlooked when self-enforcing contracts are discussed in the setting described in Chapters 3, 4 and 5. These chapters will focus on the second restriction.

2.7 SUMMARY: THE BASIC RESULTS

Future responses to current actions must be credible in order to support cooperation. One response scheme is the so-called trigger strategy. This strategy specifies that an agent reverts to infinitely lasting non-cooperation when the other agent has defected from cooperation. This is a severe threat but its credibility may be in doubt because it specifies that cooperation gains will never again be realized. After defection, the situation is the same as before cooperation had begun. Hence, both agents may wish to renegotiate future actions. If renegotiation leads to resuming cooperation, however, the threat of infinite punishment is not credible.

In general, the credibility of future responses depends on the influence of past actions on future behaviour. The concept of renegotiation-proofness reconciles the need of future responses to non-cooperative behaviour and the option to return to cooperation. A renegotiation-proof strategy is the following: act cooperatively as long as your opponent also acts cooperatively. If he/she defects from cooperation, act non-cooperatively as long as your opponent has acted cooperatively unilaterally, after which you return to cooperation. This strategy is credible if defection does not pay for your opponent. Defection does not pay if it is not profitable

- to defect in one period and to act non-cooperatively later,
- to defect in one period and to act cooperatively unilaterally during the punishment phase, and
- to refrain from acting cooperatively unilaterally in order to restart cooperation after defection.

It can be shown that the first condition is met if the two other conditions are fulfilled. When developing a strategic bargaining model for self-enforcing contracts, every contract should be renegotiation proof because bargaining should include the possibility of renegotiation. Contracts are consistent declarations of intent, and self-enforcing contracts are contracts which guarantee compliance of the contractors by their own means. In general, a contract contains two consistent declarations of intent. It contains at least two paragraphs: the first one determines the inputs of every party; the second one indicates what to do in the case of non-compliance of a party. A self-enforcing contract specifies a renegotiation-proof plan as the second paragraph, as outlined above. Because contracts serve for strategy specification, they can be expected to leave no mutual improvement unexploited. Hence, the candidates for a self-enforcing contract are those agreements which meet the conditions of renegotiation-proofness and the efficiency condition that no agent can be made better off by an alternative agreement.

But even if only efficient agreements were considered, a selection problem would arise because the set of efficient agreements is not necessarily a singleton. It is this selection problem for which the strategic bargaining approach is introduced. In the literature, strategic bargaining models have never been applied to cooperation in a non-cooperative environment. Since both approaches employ discount factors, a strategic bargaining model need not make further assumptions about the payoffs other than those which were already made for studying the scope for cooperation.

Strategic bargaining models assume a certain mover structure for the bargaining process. Suppose that two agents i and j have to split a pie of size one unanimously. After they have agreed on a certain split, they eat their share of the pie, and consumption equals current utility. Hence, the whole pie is worth the discount factor δ_i (δ_j) for agent i (j) if he/she receives the whole pie in the next period. Assume that there are three bargaining stages between which one unit of time elapses. Agent i makes a proposal in the first stage which is either accepted or rejected by agent j. If accepted, both receive their shares according to agent i's proposal, if rejected, both have to wait for the second stage in which it is up to agent j to make a proposal. In the third stage, agent i makes a proposal and j either accepts or rejects it but is not able to make a counterproposal one period later.

This simple bargaining game has a unique solution which can be developed in the backward induction fashion. In the third period, it is obvious that agent i is able to make a take-it-or-leave-it proposal. Because agent j may only accept or reject, agent i will make him indifferent between both options by offering a partition which gives agent i the whole pie. This result is anticipated by agent j for the second period when it is up to him/her to make a proposal. Hence, agent j would propose δ_i to agent i so that agent i is indifferent between receiving δ_i today and the whole pie tomorrow. Agent j would be left with $1 - \delta_i$ in the second period. Again, this result is anticipated by agent i who offers $\delta_j(1 - \delta_i)$ to agent j for the first period. Then, the split $[1 - \delta_j(1 - \delta_i), \delta_j(1 - \delta_i)]$ gives the subgame-perfect bargaining solution.

This result demonstrates that the distribution of the pie depends on the discount factors and on the specification of the first mover and the last mover. The relatively higher an agent's discount factor is, the larger is his/her share of the pie. Two generalizations make the result independent of the mover structure. First, the bargaining process may have no definite end but one may assume that there is always a chance of a counter offer. This assumption removes the advantage of a last mover because no agent will be able to make a take-it-or-leave-it proposal. Second, bargaining is a process of quick responses so that the time elapsing between offers and counter offers can be assumed to be negligible. Short bargaining periods eliminate the first-mover advantage because the delay between the first and second offer is brief. When assuming general well-behaved payoff functions Π_i and Π_j for both agents, the general solution was given by (2.52) in Section 2.5:

$$\frac{d\Pi_i/\Pi_i(\cdot)}{d\Pi_j/\Pi_j(\cdot)} = -\frac{\ln\delta_i}{\ln\delta_j}. \tag{2.52}$$

Condition (2.52) gives the interior solution for a bargaining problem defined by two general utility functions, two discount factors and infinitely brief bargaining periods. As the payoffs of the non-cooperative equilibrium are fixed, the marginal bargaining gains are equal to the marginal payoffs, that is, $dU_i = d\Pi_i$, $dU_j = d\Pi_j$.

Condition (2.52) is the general solution for a strategic bargaining game for a unique realization. Strategic bargaining for a self-enforcing contract, however, means bargaining for an infinite stream of future realizations. Whereas in conventional bargaining games an agent proposes an instantaneous realization, bargaining for a self-enforcing contract means proposing an instantaneous realization, a realization one period later, a realization two periods later and so on. In both cases, a proposal will be submitted for the earliest realizations possible at the time of the proposal. Conventional bargaining models cover only one realization, bargaining for a long-term relationship covers several periods and bargaining for a self-enforcing contract covers an infinite stream of realizations.

Obviously, a proposal submitted at the beginning of the game may comprise a dynamic path which specifies that the payoffs realized in different periods differ. Suppose that both agents have agreed upon a dynamic path which gives agent i (j) lower payoffs in earlier (later) periods and agent j (i) higher payoffs in earlier (later) periods. Such a dynamic path is efficient if agent i (j) is more patient (impatient) than agent j (i). Let the discounted sum of bargaining gains due to future realizations be denoted by $\hat{U}_i^t$ and $\hat{U}_j^t$. The superscript denotes the period of the next realization. Hence, a dynamic path proposed before the earliest realization in period 0 can occur gives $\hat{U}_i^0$ and $\hat{U}_j^0$.

A dynamic path, however, does not qualify for time-consistent bargaining behaviour. In a non-cooperative environment, every agent may submit a new proposal which replaces the old agreement whenever he/she wants to. Additionally, agreements reflect the bargaining power of both agents in this setting. If the dynamic path implied $\hat{U}_i^0$ and $\hat{U}_j^0$ before period 0, this result is due to the bargaining power of both agents. When the first realization has occurred according to the proposal, however, $\hat{U}_i^1$ and $\hat{U}_j^1$ according to this proposal no longer reflect bargaining power: as the dynamic path gives agent i a lower utility in the beginning and higher utilities in later periods, $\hat{U}_i^1 > \hat{U}_i^0$ and $\hat{U}_j^1 < \hat{U}_j^0$ hold. Hence, the bargaining result will be questioned by agent j after the first realization has occurred. He/she will argue that the situation is the same as before the first realization. The dynamic path is therefore not credible unless one makes the unrealistic assumption that both agents can credibly commit themselves never to question an agreement by submitting a new proposal.

Time-consistent bargaining behaviour requires that a self-enforcing contract will never be subject to revision in the future. In this setting, only stationary contracts which imply time-independent payoffs for each agent in every period fulfil this requirement. As both agents anticipate that only stationary contracts guarantee time-consistent bargaining behaviour, proposals will be restricted to stationary proposals. Then, it can clearly be seen that (2.52) gives the solution for strategic bargaining for a self-enforcing contract as well. For stationary contracts, the sum of discounted present and future payoffs is $1/(1 - \delta_i)$ or $1/(1 - \delta_j)$, respectively, times the current payoffs, so that (2.52) depends only on the current payoffs because $1/(1 - \delta_i)$ and $1/(1 - \delta_j)$ both cancel out when dividing the respective total differential $d\Pi$ by Π. Thus, strategic bargaining for a self-enforcing contract yields the same result as strategic bargaining for one realization. Chapters 3, 4 and 5 may therefore employ (2.52) to determine the strategic bargaining solution.

A certain point should be stressed in order to avoid confusion. The general time structure of the model is not repetition of realizations but the assumption that the opportunities for realizations are repeated. Realization may be delayed because of a disagreement in bargaining, and a subsequent realization is only possible after one unit of time has elapsed from the previous factual realization. Time-consistent bargaining behaviour implies factual repetition, and delays never occur because this would imply delay costs for both agents. Hence, the general model is not restricted to a repeated game but results in factual repetition.

All models in this book assume two agents i and j, whose payoffs depend on both agents' strategies. If they do not coordinate their actions, they are worse off than if they do, but every agent is even better off if the other agent cooperates but he/she does not. The game is a game of almost perfect information so that all is common knowledge but both agents have to move simultaneously when deciding on their individual actions, without being able to observe the other agent's actions. Hence, both agents are involved in a typical prisoners' dilemma game for which only non-cooperation of both agents constitutes an equilibrium in a one-shot game. Chapters 3, 4 and 5 will discuss strategic bargaining for cooperation in different settings when this one-shot game is repeated or is likely to be repeated. Table 2.4 summarizes the assumptions and their implications.

Table 2.4 Model overview

Assumption	Implication
Unique non-cooperative equilibrium is Pareto inferior	Bargaining gains may be realized
Potentially infinite repetition	Cooperation may be sustained by repetition, no change of the environment
Fixed discount factors reflect impatience (institutional toughness)	Finite and infinite sums of geometric series give the discounted sum of payoffs
Self-enforcing contracts must be weakly renegotiation proof	Two relevant compliance constraints for each agent
Time to elapse between two realizations is at least one	Delay may occur due to disagreement
Time-consistent bargaining behaviour for long-run cooperation	Self-enforcing contracts are stationary
Set of efficient agreements gives a concave and at least piecewise differentiable frontier	Uniqueness of the strategic bargaining solution
Strategic bargaining in an alternating offer game with infinitely small bargaining periods and no termination	Elimination of first- and last-mover advantages

3. Bargaining for a Collective Good without Transfers

3.1 INTRODUCTION: SELF-ENFORCING COOPERATION WITHOUT TRANSFERS

When considering bargaining between two agents for the joint provision or joint production of a collective good, two cases have to be distinguished. The distinction originates from the institutional setting which either supports or fights against the policies of two agents who aim at an agreement which Pareto dominates the non-cooperative outcome. On the one hand, cooperation between two agents may be appreciated by third parties. In this case, the two agents will be given the maximum freedom with respect to instruments to be used in order to reach the maximum degree of cooperation. In particular, the two agents will be allowed to use transfers from one agent to the other which serve to exploit mutual bargaining gains when individual benefits and costs differ substantially. For example, any success of an international environmental agreement will obviously be appreciated and cooperation may be supported by transfers from one country to the other.

On the other hand, cooperation between two agents might decrease the utility of third parties. In this case, regulating institutions aim at hindering cooperation and will not allow open transfers as a means of cooperation policy. For example, collusion between two suppliers in an industrial market is likely to affect the utility of third parties, for example, consumers, negatively. If transfers which are not market based because they merely redistribute collusion gains can be verified and prohibited by regulating authorities, cooperation will be possible only without transfers.

This chapter deals with the second case. It assumes that cooperation between the two agents will be hindered by the institutional setting in addition to necessary self-enforcement. This setting enables a regulating authority to verify and prohibit transfers which are paid to redistribute cooperation gains. Auditing by a board is assumed to be able to distinguish between transfers paid for exchanges in the market place and transfers paid to distribute gains from implicit collusion. However, the regulating authorities are assumed to be unable to verify whether other actions are caused by non-cooperative or collusive behaviour. Chapter 4 will deal with bargaining for a collective good if transfers are allowed.

The most prominent application of bargaining for a collective good without transfers is implicit or 'tacit' collusion between two duopolists. The notion of tacit collusion will not be used here as tacit would be a misleading term when communication plays a crucial role for bargaining between two duopolists. There is some ambiguity in the use of these terms. For example, Rees (1993) deals with an agreement on coordinated market behaviour which is due to communication among oligopolists as an explicit agreement. His notion of implicit collusion describes market behaviour without any coordination which does not produce the competitive solution. In the sense of this book, this is an important difference because Rees's notion of implicit collusion corresponds to the notion of repeated games and his notion of explicit collusion which must be self-enforcing corresponds to the notion of self-enforcing contracts. The notion of implicit collusion as it is used here, however, should indicate that communication among oligopolists has led to a solution which gives at least one oligopolist higher profits but cannot be verified by an antitrust board.

The literature on possible collusion in industrial markets has been very extensive since the Folk Theorem was formulated. Collusion among oligopolists represents the prototype of applications which encouraged game theorists to consider behaviour and equilibria in repeated games. Every standard industrial organization textbook, such as Martin (1993), Tirole (1988) and Scherer and Ross (1990), covers collusive aspects. But collusion is often understood as aiming at monopolistic behaviour. A firm aiming at coordination with one competitor, however, is not interested in sustaining the monopolistic outcome but in maximizing its profits subject to the compliance of the other firm. The monopolistic outcome would define the duopolists' best policy if and only if transfers were possible in order to shift collusive gains and transfers were not at risk of non-compliance. Both conditions are ruled out here because transfers are excluded by regulation. The next chapter will allow for transfers but transfers will be subject to compliance risks as well.

This chapter also assumes that both agents under consideration are selfish and interested not in maximizing joint profits but individual profits. There is no binary choice of actions to be taken but options are many and they are therefore subject to bargaining. As the model will show that results which reflect joint profit maximization are not guaranteed, it may explain performances of imperfect collusion which mirror neither monopolistic behaviour nor competition (even if the monopolistic outcome could be sustained on the basis of repetition). The basic model of this chapter specifies benefit and cost functions which imply orthogonal reaction functions. Orthogonal reaction functions are used in order to facilitate developing compliance constraints and a concession line which includes all candidates for the bargaining solution. The basic model is the limit case of strategic complements. In a general setting, the set of compliance constraints determining one part of the concession line would be doubled. Orthogonal reaction functions, however, are at the least a rather strange assumption for duopolistic markets. But the model which will employ orthogonal reaction functions will restrict complexity significantly and will show effects which hold in general. Section 3.5 will demonstrate that the results carry over to non-orthogonal reaction functions. This section will also show some results for a duopolistic market for strategic substitutes.

Coordination on duopolistic market performance is not the only conceivable application for two firms. Another application may be found in providing a collective good for two firms producing not for a common market but for different markets which depend on each other. For example, a software and a hardware company may increase their profits if both advertising campaigns stress the compatibility with the other firm's product. However, each firm is even better off if it is promoted by the other firm without stressing the other firm's compatibility with its product: the software (hardware) company would enjoy the recommendation by the hardware (software) company but would not restrict its market to the other company's hardware (software) users. Thus, both companies have an incentive to break any implicit agreement on mutual recommendations. This breach option can be ruled out by a self-enforcing contract specifying advertising programmes for both firms which guarantee compliance. Coordinated advertising is not the only application for firms which operate in different but interdependent markets. Other applications are coordination on compatibility standards or on joint distribution. In the first case, two firms agree upon a certain (not necessarily efficient) compatibility standard which makes their goods strongly complementary; in the second case, they agree upon distributing their goods only to a limited number of retailers in order to increase the probability that consumers buy both products. Both cases give typical examples for the provision of a collective good; the enforceability of both by binding contracts is likely to be ruled out by antitrust policies.

This chapter is organized as follows. Section 3.2 introduces the model assumptions and develops the compliance constraints that weak renegotiation-proofness imposes. Additionally, it prepares strategic bargaining between the two agents by specifying the optimal policies for both agents. Section 3.3 introduces the concession line which defines the set of efficient agreements. Each element of this set is a potential candidate for the strategic bargaining solution. Section 3.3 also discusses some properties of the concession line which depend on the discount factors of both agents. After developing the concession line, determining the strategic bargaining solution will be straightforward. Section 3.4 deals with the role of discount factors for the strategic bargaining solution. It demonstrates that a marginal change in an agent's discount factor may increase or decrease this agent's utility. This change will be shown to depend on the specific range of the concession line which 'hosts' the strategic bargaining solution. Section 3.5 gives an application of strategic bargaining for self-enforcing contracts without transfers. It discusses implicit collusion in a duopolistic market. Section 3.6 summarizes the results of this chapter.

Compared to the following chapters, this chapter will be more explicit about developing the model. This explicit formulation opens the way for an analysis of the differences between the models of the following chapters and this model to be made in the rest of the book. As the basic assumptions are very similar, the extensions and modifications dealt with in Chapters 4 and 5 will then be clear.

3.2 MODEL ASSUMPTIONS, COMPLIANCE CONSTRAINTS AND OPTIMAL POLICIES

The theoretical model of this chapter assumes that two agents i and j produce a collective good through individual efforts. In order to distinguish the specific models from the general model, the notation is changed. The gross benefits B_i and B_j are linear with respect to the sum of efforts and the costs C_i and C_j are quadratic with respect to the individual efforts E_i and E_j:

$$B_i = \alpha_i\left(E_i + E_j\right), \ C_i = \frac{\gamma_i}{2}E_i^2, \ V_i = B_i - C_i, \ \alpha_i, \gamma_i > 0,$$

$$B_j = \alpha_j\left(E_i + E_j\right), \ C_j = \frac{\gamma_j}{2}E_j^2, \ V_j = B_j - C_j, \ \alpha_j, \gamma_j > 0. \ (3.1)$$

V_i and V_j mirror the utilities that agent i and agent j enjoy. Utilities are cardinal because they are measured in payoffs. The functions denoted by V are therefore payoff functions in the sense of (2.12) (which were denoted by Π in Chapter 2). The linearity of benefits will simplify the analysis significantly because it induces orthogonal reaction functions. The non-cooperative Nash equilibrium defines the payoff both agents receive in any case; the difference between the realized payoffs and the non-cooperative payoff level gives the bargaining gains. According to (3.1),

$$E_i^n = \frac{\alpha_i}{\gamma_i}, \quad E_j^n = \frac{\alpha_j}{\gamma_j} \tag{3.2}$$

are the non-cooperative effort levels which are denoted by the superscript n. Equation (3.2) also gives the outside option of each agent. Orthogonality of reaction functions implies that a non-compliant agent takes the free ride without adjusting to a lower effort level. Consider, for example, agent i and assume that agent i intends to break an agreement for which agent j has to carry E_j. The outside option of agent i is the solution of

$$\max_{E_i} \left\{ \alpha_i \left(E_i + E_j \right) - \frac{\gamma_i}{2} E_i^2 \right\}$$

which is the same problem as in the case of non-cooperative policies. The efforts of agent i if agent i chooses the outside option (which will also be referred to as the breach level) do not depend on E_j and are given by the first term of (3.2).

If stationary contracts were completely enforceable, both agents could agree upon the cooperative effort levels

$$E_i^* = \frac{\alpha_i + \alpha_j}{\gamma_i}, \quad E_j^* = \frac{\alpha_i + \alpha_j}{\gamma_j}. \tag{3.3}$$

Equation (3.3) maximizes the total utilities, and (3.4) gives the individual net utilities U_i^* and U_j^* as the difference between the cooperative utilities V_i^*, V_j^* and the non-cooperative utilities V_i^n, V_j^n:

$$U_i^* = V_i^* - V_i^n = \frac{\alpha_i^2}{\gamma_j} - \frac{\alpha_j^2}{2\gamma_i}.$$

$$U_j^* = V_j^* - V_j^n = \frac{\alpha_j^2}{\gamma_i} - \frac{\alpha_i^2}{2\gamma_j}. \tag{3.4}$$

Note that the individual net utilities according to (3.4) must be positive to imply a pure collective good problem. Otherwise, one agent would have to carry net costs from cooperation if he/she received no transfers and the problem would change partially into a bargaining problem between a buyer and a seller. In this case, the cooperative effort levels according to (3.4) are not feasible without transfers. Chapter 4 will deal explicitly with this case.

The total bargaining gains U of cooperation if contracts were enforceable sum up to

$$U = \sum_{k=i,j} \left[V_k^* - V_k^n \right] = \frac{\alpha_i^2}{2\gamma_j} + \frac{\alpha_j^2}{2\gamma_i}. \tag{3.5}$$

In the case of enforceable contracts, both agents would maximize their individual utility by an effort level according to (3.3) which maximizes the total bargaining gains because effort allocation and distribution of U can be strictly separated through transfers. As both agents receive a positive share whose size depends on their relative impatience, they are both interested in realizing the largest possible total bargaining gains. Then, the individual efforts affect only the determination of U but not its distribution, which is determined by the solution of the strategic bargaining game.

Compliance Constraints and Utility Reservation Functions

Chapter 2 has demonstrated that self-enforcing contracts should meet the conditions of weak renegotiation-proofness according to Farrell and Maskin (1989). A weakly renegotiation-proof plan proposed for Paragraph 2 of the self-enforcing contract was the rule that an agent who has broken the contract must provide the agreed-upon effort level unilaterally in n periods in order to resume cooperation in period $(n + 1)$. This rule specifies three constraints which every agreement must fulfil to qualify for a self-enforcing contract. Note that all constraints are given in terms of the benefits V_i and V_j whereas Chapter 2 employed bargaining gains. The use of V_i and V_j simplifies the development of critical effort levels in this case because several terms are identical due to the orthogonality of the reaction functions. The first constraint is well known from infinitely repeated games which are subject to trigger strategies. Condition (3.6) requires that every self-enforcing contract must guarantee that the benefits from fulfilling the contract must not fall short of the benefits from breaking the contract in one period and no cooperation in all following periods.

$$\frac{1}{1-\delta_i}\left[\alpha_i\left(E_i+E_j\right)-\frac{\gamma_i}{2}E_i^2\right]-\left[\frac{\alpha_i^2}{2\gamma_i}+\alpha_iE_j\right]-\frac{\delta_i}{1-\delta_i}\left[\frac{\alpha_i^2}{2\gamma_i}+\frac{\alpha_i\alpha_j}{\gamma_j}\right]\geq0,$$

$$\frac{1}{1-\delta_j}\left[\alpha_j\left(E_i+E_j\right)-\frac{\gamma_j}{2}E_j^2\right]-\left[\frac{\alpha_j^2}{2\gamma_j}+\alpha_jE_i\right]-\frac{\delta_j}{1-\delta_j}\left[\frac{\alpha_j^2}{2\gamma_j}+\frac{\alpha_i\alpha_j}{\gamma_i}\right]\geq0.$$

$$(3.6)$$

Equation (3.6) is independent of n, and it gives only a necessary but not sufficient condition for a self-enforcing contract. The first term mirrors the sum of discounted benefits which arise from infinitely repeated cooperation. This sum should not fall short of the benefits of non-compliance (second term) and the benefits which arise from infinite non-cooperation started one period later (third term). The second and the third term show that the non-cooperative level and the breach level fall together because of the orthogonality of the reaction functions.

Weak renegotiation-proofness was shown to add another two conditions which depend on n. First, every self-enforcing contract must meet the condition of *ex ante* compliance. An agent should not be better off by breaking the contract in one period and investing in a resumption of cooperation in n, $n \geq 1$, periods compared to compliance in $(n + 1)$ periods:

$$\frac{1-\delta_i^{n+1}}{1-\delta_i}\left\{\alpha_i\left[E_i+E_j\right]-\frac{\gamma_i}{2}E_i^2\right\}-\left[\frac{\alpha_i^2}{2\gamma_i}+\alpha_iE_j\right]$$

$$-\delta_i\frac{1-\delta_i^n}{1-\delta_i}\left\{\alpha_i\left[E_i+\frac{\alpha_j}{\gamma_j}\right]-\frac{\gamma_i}{2}E_i^2\right\}\geq0,$$

$$\frac{1-\delta_j^{n+1}}{1-\delta_j}\left\{\alpha_j\left[E_i+E_j\right]-\frac{\gamma_j}{2}E_j^2\right\}-\left[\frac{\alpha_j^2}{2\gamma_j}+\alpha_jE_i\right]$$

$$-\delta_j\frac{1-\delta_j^n}{1-\delta_j}\left\{\alpha_j\left[E_j+\frac{\alpha_i}{\gamma_i}\right]-\frac{\gamma_j}{2}E_j^2\right\}\geq0. \qquad (3.7)$$

Second, every self-enforcing contract should meet the condition of *ex post* compliance. After an agent has broken the contract, he/she should be weakly better off by investing n periods in a resumption of cooperation compared to no cooperation in all future periods. Condition (3.8) requires that the benefits from a resumption of cooperation (second term) minus the costs of carrying the agreed-upon effort level in n periods (first term which falls short of the non-cooperative utility levels) must not fall short of the benefits of no cooperation in all future periods (third term).

$$\frac{1-\delta_i^n}{1-\delta_i}\left\{\alpha_i\left[E_i+\frac{\alpha_j}{\gamma_j}\right]-\frac{\gamma_i}{2}E_i^2\right\}+\frac{\delta_i^n}{1-\delta_i}\left\{\alpha_i\left[E_i+E_j\right]-\frac{\gamma_i}{2}E_i^2\right\}$$

$$-\frac{1}{1-\delta_i}\left[\frac{\alpha_i^2}{2\gamma_i}+\frac{\alpha_i\alpha_j}{\gamma_j}\right]\geq 0,$$

$$\frac{1-\delta_j^n}{1-\delta_j}\left\{\alpha_j\left[E_j+\frac{\alpha_i}{\gamma_i}\right]-\frac{\gamma_j}{2}E_j^2\right\}+\frac{\delta_j^n}{1-\delta_j}\left\{\alpha_j\left[E_i+E_j\right]-\frac{\gamma_j}{2}E_j^2\right\}$$

$$-\frac{1}{1-\delta_j}\left[\frac{\alpha_j^2}{2\gamma_j}+\frac{\alpha_i\alpha_j}{\gamma_i}\right]\geq 0. \tag{3.8}$$

Conditions (3.6), (3.7) and (3.8) have been derived in their general form in Section 2.4 (see (2.16), (2.17) and (2.19)). Which n maximizes the scope for cooperation sustained by a self-enforcing contract? Assume that n is set equal to 2 and consider the constraints which are determined by potential deviations of agent i. For $n = 2$, the first line of (3.7) which specifies *ex ante* compliance can be rewritten as a minimum effort level condition for agent j. Agent j's efforts must fulfil (3.9) in order to guarantee *ex ante* compliance of agent i:

$$E_j\geq\frac{\alpha_j}{\gamma_j}+\frac{1}{\delta_i+\delta_i^2}\frac{\gamma_i}{2\alpha_i}\varepsilon_i^2. \tag{3.9}$$

According to (3.8), *ex post* compliance requires

$$E_j\geq\frac{\alpha_j}{\gamma_j}+\frac{1}{\delta_i^2}\frac{\gamma_i}{2\alpha_i}\varepsilon_i^2, \tag{3.10}$$

and (3.6) requires as a general condition for cooperation

$$E_j\geq\frac{\alpha_j}{\gamma_j}+\frac{1}{\delta_i}\frac{\gamma_i}{2\alpha_i}\varepsilon_i^2. \tag{3.11}$$

In (3.9), (3.10) and (3.11), $E_i - \alpha_i/\gamma_i$ has been replaced by ε_i. In the rest, ε_i and ε_j will denote the degree of effort above the non-cooperative level. Every self-enforcing contract must meet all three conditions which define the minimum efforts of j and the corresponding conditions for i if n is set equal to two. Note that all conditions give the non-cooperative level if the other agent is supposed to realize his/her non-cooperative level, too. If agent i is supposed to increase his/her efforts, agent j must provide a specific effort level in order to make non-compliance for the other agent unattractive. This level depends crucially on agent j's discount factor which determines the degree to which future punishments are taken into account for current actions. Given the effort level of agent i, the more impatient agent i is, the more efforts must be provided by agent j in order to guarantee agent i's compliance. This result holds generally independent of the chosen n. For $n = 2$, it is obvious that (3.10) dominates (3.9) and (3.11) because $1/\delta_i^2$ exceeds both $1/\delta_i$ and $1/(\delta_i + \delta_i^2)$. A general result that (3.11) is always met if (3.9) and (3.10) are has been derived in Section 2.4.

Considering the determination of n means asking the question whether the dominating constraint is too demanding. A too demanding constraint would restrict the scope for cooperation more than necessary if another specification allowed further beneficial outcomes to be realized. Obviously, increasing n strengthens the power of the *ex ante* compliance constraint but weakens the power of the *ex post* compliance constraint because an increase in n increases the costs for resuming cooperation. Lowering n from 2 to 1 lets all three constraints coincide so that (3.11) must hold. Therefore, setting n higher than one makes the *ex post* compliance constraint too demanding. A one-period punishment implies compliance constraints which guarantee the maximum scope for cooperation. If benefits and costs can be represented by (3.1), $n = 1$ minimizes the burden the compliance constraints impose on potential cooperation. Section 2.4 has demonstrated that this result holds for orthogonal reaction functions in general.

For $n = 1$, all three compliance constraints define a critical effort level of which the one agent's (i's or j's) efforts should not fall short in order to ensure the compliance of the other agent (j or i):

$$\varepsilon_j^{C(i)} = \frac{1}{\delta_i}\frac{\gamma_i}{2\alpha_i}\varepsilon_i^2, \quad \varepsilon_i^{C(j)} = \frac{1}{\delta_j}\frac{\gamma_j}{2\alpha_j}\varepsilon_j^2. \tag{3.12}$$

The superscripts $C(i)$ and $C(j)$ denote the compliance constraints which agent i and agent j impose on the other agent's efforts. These functions will be referred to as compliance constraints. Drawn in figures, (3.12) will be referred to as compliance constraint curves. Differentiating (3.12) gives the slopes of the compliance constraints:

$$\left(\frac{dE_j}{dE_i}\right)^{C(i)} = \frac{1}{\delta_i}\left[\frac{\gamma_i}{\alpha_i}E_i - 1\right], \quad \left(\frac{dE_i}{dE_j}\right)^{C(j)} = \frac{1}{\delta_j}\left[\frac{\gamma_j}{\alpha_j}E_j - 1\right]. \tag{3.13}$$

It is worthwhile to compare the compliance constraints with the utility reservation functions. Let $R(i)$ and $R(j)$ indicate the effort levels which give the non-cooperative utility to agent i and agent j, respectively. The effort vectors which define indifference to the non-cooperative utilities can be derived from

$$V_i\left[E_i^{R(i)}, E_j^{R(i)}\right] = \alpha_i\left[E_i^{R(i)} + E_j^{R(i)}\right] - \frac{\gamma_i}{2}E_i^{R(i)^2} = \frac{\alpha_i^2}{2\gamma_i} + \frac{\alpha_i\alpha_j}{\gamma_j},$$

$$V_j\left[E_i^{R(j)}, E_j^{R(j)}\right] = \alpha_j\left[E_i^{R(j)} + E_j^{R(j)}\right] - \frac{\gamma_j}{2}E_j^{R(j)^2} = \frac{\alpha_j^2}{2\gamma_j} + \frac{\alpha_i\alpha_j}{\gamma_i}. \tag{3.14}$$

$[E_i^{R(i)}, E_j^{R(i)}]$ describes the effort vector which leaves agent i on his/her non-cooperative utility level, $[E_i^{R(j)}, E_j^{R(j)}]$ describes the effort vector which leaves agent j on his/her non-cooperative utility level. Equation (3.14) can be used to derive the utility reservation functions

$$E_j^{R(i)} = \frac{\gamma_i}{2\alpha_i}\varepsilon_i^2, \quad E_i^{R(j)} = \frac{\gamma_j}{2\alpha_j}\varepsilon_j^2 \tag{3.15}$$

which give the necessary effort level of one agent in order to leave the other agent on his/her non-cooperative utility level. The slopes of these functions are given by

$$\left(\frac{dE_j}{dE_i}\right)^{R(i)} = \frac{\gamma_i}{\alpha_i}E_i - 1,$$

$$\left(\frac{dE_i}{dE_j}\right)^{R(j)} = \frac{\gamma_j}{\alpha_j}E_j - 1. \tag{3.16}$$

As the non-cooperative utility level is a constant in (3.14), the RHS of (3.16) not only gives the slope of the utility reservation curve but of any indifference curve as well. Thus, (3.16) also holds for every iso-utility curve. Comparing (3.13) and (3.16) shows that the compliance constraints (3.12) have a steeper slope than the utility reservation curves (3.15). The utility reservation function is a specific iso-utility curve and the limit of the compliance constraint (3.12) when $\delta_i \to 1$ or $\delta_j \to 1$, respectively.

Optimal Policies and the Feasibility of the Contract Curve

One may now consider the impact of the compliance constraints on the possible solutions, especially on the optimal policies of an agent. Figure 3.1 gives the utility reservation curve of agent i (see the first line of (3.15)), the compliance constraint defined by agent i (see the first line of (3.13)) and two iso-utility curves of agent j.

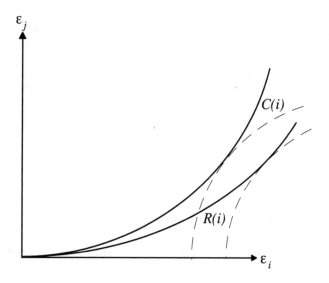

Figure 3.1   Utility reservation and compliance constraint of agent i

In Figure 3.1, the intersection of the axes gives not the zero efforts but the non-cooperative effort levels. $R(i)$ denotes the utility reservation curve and $C(i)$ denotes the compliance constraint which has a steeper slope than $R(i)$. The iso-utility curves of agent j imply a utility increase when they are shifted rightwards because an increase in E_i for a constant E_j benefits agent j (see (3.1)). Figure 3.1 shows that the compliance constraint changes the optimal policy of agent j: the tangential point with the utility reservation curve is not a credible agreement but ensures the defection of agent *i*. The optimal policy for *j* makes agent *i* indifferent between compliance and non-compliance and gives him/her strictly positive bargaining gains.

Figure 3.2 gives the respective picture for agent *j*.

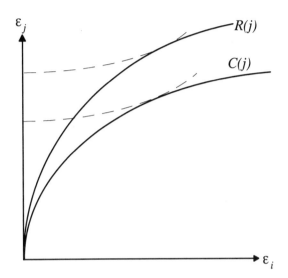

Figure 3.2   Utility reservation and compliance constraint of agent j

Figure 3.2 is a mirror image of Figure 3.1. The steeper slope of the compliance constraint *C(j)* must be read on the basis of this mirror image because *C(j)* specifies minimum efforts of agent *i* which are to be read on the abscissa. The indifference curves of agent *i* imply a utility increase when they are shifted upwards because an increase in E_j for a constant E_i benefits agent *i*.

Figures 3.1 and 3.2 have assumed sufficiently high discount factors which imply compliance constraints in the close neighbourhood of the utility reservation curves. Figure 3.3 shows the scope for cooperation if both figures are combined for such a scenario. The optimal policy of agent *i* (*j*) for an enforceable contract is given by A (B), the optimal policy for agent *i* (*j*) for a self-enforcing contract is given by C (F). The dotted curves give indifference curves which demonstrate the set of efficient agreements in the line CDEF. In point 1, two indifference curves strictly in the inner lens below the contract curve intersect. The contract curve denotes the loci of effort combinations which are Pareto optimal without any compliance constraints. Point 1 gives agent *i* (*j*) the utility he/she also realizes in point F (E). Point 1 is Pareto dominated by all points on FE. In point 2 (3), two indifference curves strictly in the inner lens above (below) the contract curve intersect. Point 2 and Point 3 are both Pareto dominated by points on the contract curve.

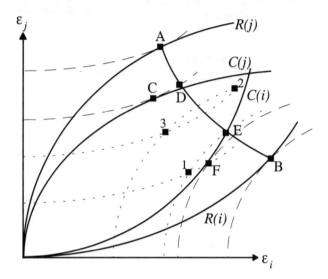

Figure 3.3   Scope for cooperation without transfers: the case of high discount factors

The vague notion of sufficiently high discount factors will be given an exact clarification later on in this section. Any improvement on the non-cooperative outcome must lie between the utility reservation curves. The line AB gives the contract curve which would be referred to as the set of Pareto-optimal outcomes if contracts were enforceable. On the contract curve,

$$dU_i = (\alpha_i - \gamma_i E_i)dE_i + \alpha_i dE_j = 0,$$
$$dU_j = (\alpha_j - \gamma_j E_j)dE_j + \alpha_j dE_i = 0 \qquad (3.17)$$

must hold to ensure Pareto optimality. Conditions (3.17) can be rewritten as differential conditions

$$\left[\frac{dE_j}{dE_i}\right]^{I_i} = -\frac{\alpha_i - \gamma_i E_i}{\alpha_i}, \quad \left[\frac{dE_j}{dE_i}\right]^{I_j} = -\frac{\alpha_j}{\alpha_j - \gamma_j E_j} \qquad (3.18)$$

The superscripts I_i and I_j denote the slope of agent i's and agent j's indifference curve on the contract curve, respectively. Equation (3.18) determines the functional form of the contract curve (denoted by the superscript CO),

$$E_j^{CO} = \frac{\alpha_j}{\gamma_j} + \frac{\alpha_i}{\gamma_i} \frac{\alpha_j}{\gamma_j \varepsilon_i^{CO}} \tag{3.19}$$

which is defined for effort levels between A and B only. The slope of (3.19) is given by

$$\frac{dE_j^{CO}}{dE_i^{CO}} = -\frac{\alpha_i}{\gamma_i} \frac{\alpha_j}{\gamma_j \varepsilon_i^{CO2}} < 0,$$

$$\frac{d^2 E_j^{CO}}{dE_i^{CO2}} = \frac{2\alpha_i}{\gamma_i \gamma_j} \frac{\alpha_j}{\varepsilon_i^{CO3}} > 0. \tag{3.20}$$

Figure 3.3 shows how the compliance constraints restrict the solution set. Solutions have to be found in the inner lens embraced by the compliance constraints. Section 2.5 has shown that no strategic bargaining solution will leave any bargaining gains unexploited. Hence, strategic bargaining solutions will be restricted to the contract curve and parts of the compliance constraints. These parts are given by the arcs FE and CD in Figure 3.3. On these arcs and on the contract curve, no agent can be made better off without making the other agent worse off. All other solutions leave scope for improvement.

The compliance constraints restrict the range for Pareto improvements. The lower the discount factor of one agent, the higher must be the critical effort level of the other agent. Figure 3.3 shows a bargaining problem which does not dominate the whole contract curve AB. The branch DE does not violate the compliance constraints. At first glance, one could expect that both agents will agree upon the optimal effort vector which maximizes their joint benefits subject to the compliance constraints. When the allocation of efforts and the distribution of bargaining gains cannot be decoupled, however, the determination of the outcome is much more complex because the effort vector itself is subject to strategic bargaining. The size of the total bargaining gains becomes irrelevant because the agreed-upon efforts determine directly the individual utilities of both agents.

According to Figure 3.3, agent i (j) increases his/her utility by shifting his/her iso-utility curve to the northwest (southeast). A (B) gives the point on which i's (j's) utility curve is tangential to j's (i's) utility reservation curve whereas C (D) reflects the point on which i's (j's) iso-utility curve is tangential to the compliance constraint defined by j (i). In a setting which guarantees enforceable contracts but no transfers, A (B) represents the effort vector which leaves the utility of j (i) unchanged and enables i (j) to reap all of the bargaining gains. Obviously, the need to make contracts self-enforcing shrinks the potential to reap the whole bargaining gains because C (F) implies an improvement for j (i) although it represents the best possible policy for i (j) (see Figures 3.1 and 3.2).

Computing the effort vectors at these points is straightforward because the slopes of the indifference curves and the slope of the compliance constraints allow the optimal policies to be determined by tangential curves. At point C (F), the indicated iso-utility curve of i (j) is tangential to the compliance constraint curve. Thus, the maximization problem whose solution defines the optimal policies can be developed by equalizing the slope of the iso-utility curve and the slope of the compliance constraint. Consider agent j who reaches his/her highest obtainable utility as the solution of

$$\max_{E_j}\left\{\alpha_j\left[E_j + E_i\left(E_j\right)\right] - \frac{\gamma_j}{2}E_j^2\right\} \Rightarrow \alpha_j\left(1 + \frac{dE_i}{dE_j}\right) - \gamma_j E_j = 0. \qquad (3.21)$$

Since the utility reservation curves and the compliance constraints can be inverted in the relevant range, the differential can be derived by the use of the respective inverse of (3.13). For the best self-enforcing agreement in point F, inserting (3.13) into (3.21) gives a functional for E_j which must be equal to the critical E_j developed by (3.12) in order to guarantee the best possible policy for j:

$$E_j = \frac{\alpha_j}{\gamma_j} + \frac{\alpha_j}{\gamma_j}\delta_i\frac{\alpha_i}{\gamma_i E_i - \alpha_i} \stackrel{!}{=} \frac{\alpha_j}{\gamma_j} + \frac{1}{\delta_i}\frac{\gamma_i}{2\alpha_i}\left[E_i - \frac{\alpha_i}{\gamma_i}\right]^2. \qquad (3.22)$$

Equation (3.22) allows the effort vector for point F to be determined:

$$E_i^F = \sqrt[3]{\delta_i^2\frac{2\alpha_i^2\alpha_j}{\gamma_i^2\gamma_j}} + \frac{\alpha_i}{\gamma_i}, \quad E_j^F = \frac{1}{\delta_i}\frac{\gamma_i}{2\alpha_i}\left[\delta_i^2\frac{2\alpha_i^2\alpha_j}{\gamma_i^2\gamma_j}\right]^{2/3} + \frac{\alpha_j}{\gamma_j}. \qquad (3.23)$$

Since the slope of the utility reservation curve and the minimal E_j which leaves i's utility unchanged for changing efforts of i are known, the effort vector for point B can be derived in the same way:

$$E_i^B = 3\sqrt{\frac{2\alpha_i^2\alpha_j}{\gamma_i^2\gamma_j}} + \frac{\alpha_i}{\gamma_i}, \quad E_j^B = \frac{\gamma_i}{2\alpha_i}\left[\frac{2\alpha_i^2\alpha_j}{\gamma_i^2\gamma_j}\right]^{2/3} + \frac{\alpha_j}{\gamma_j}. \tag{3.24}$$

Similarly, A and C are determined by:

$$E_i^C = \frac{1}{\delta_j}\frac{\gamma_j}{2\alpha_j}\left[\delta_j^2\frac{2\alpha_j^2\alpha_i}{\gamma_j^2\gamma_i}\right]^{2/3} + \frac{\alpha_i}{\gamma_i}, \quad E_j^C = 3\sqrt{\delta_j^2\frac{2\alpha_j^2\alpha_i}{\gamma_j^2\gamma_i}} + \frac{\alpha_j}{\gamma_j}, \tag{3.25}$$

$$E_i^A = \frac{\gamma_j}{2\alpha_j}\left[\frac{2\alpha_j^2\alpha_i}{\gamma_j^2\gamma_i}\right]^{2/3} + \frac{\alpha_i}{\gamma_i}, \quad E_j^A = 3\sqrt{\frac{2\alpha_j^2\alpha_i}{\gamma_j^2\gamma_i}} + \frac{\alpha_j}{\gamma_j}. \tag{3.26}$$

Figure 3.3 shows a coordination problem which does not dominate the whole contract curve. Equalizing (3.12) and (3.19) allows the effort vectors at point D and point E to be determined:

$$E_i^D = \frac{1}{\delta_j}\frac{\gamma_j}{2\alpha_j}\left[\delta_j\frac{2\alpha_j^2\alpha_i}{\gamma_j^2\gamma_i}\right]^{2/3} + \frac{\alpha_i}{\gamma_i}, \quad E_j^D = 3\sqrt{\delta_j\frac{2\alpha_j^2\alpha_i}{\gamma_j^2\gamma_i}} + \frac{\alpha_j}{\gamma_j}, \tag{3.27}$$

$$E_i^E = 3\sqrt{\delta_i\frac{2\alpha_i^2\alpha_j}{\gamma_i^2\gamma_j}} + \frac{\alpha_i}{\gamma_i}, \quad E_j^E = \frac{1}{\delta_i}\frac{\gamma_i}{2\alpha_i}\left[\delta_i\frac{2\alpha_i^2\alpha_j}{\gamma_i^2\gamma_j}\right]^{2/3} + \frac{\alpha_j}{\gamma_j}. \tag{3.28}$$

D and E converge to A and B when the discount factors approach one. Only if

$$E_i^D < E_i^E \Leftrightarrow E_j^D > E_j^E \tag{3.29}$$

holds, the compliance constraints do not dominate the whole contract curve. If (3.29) is not valid, no self-enforcing contract exists which is in line with conventional Paretian demands and agreements on the contract curve cannot be sustained by a self-enforcing contract. Condition (3.29) implies

$$\left[\delta_i \frac{2\alpha_i^2 \alpha_j}{\gamma_i^2 \gamma_j} \right]^{1/3} > \frac{1}{\delta_j} \frac{\gamma_j}{2\alpha_j} \left[\delta_j \frac{2\alpha_j^2 \alpha_i}{\gamma_j^2 \gamma_i} \right]^{2/3}$$

which leads to

$$\delta_i \delta_j > \frac{1}{4}. \tag{3.30}$$

If either δ_i or δ_j fall short of 0.25, the compliance constraints dominate the contract curve completely and no agreement on the contract curve is feasible. In this case, any solution must be found below the contract curve. But even if both discount factors exceed 0.25, the feasibility of self-enforcing contracts on the contract curve is not guaranteed. For example, $\delta_i = 0.4$ does not suffice to satisfy (3.30) unless δ_j exceeds 0.625. If (3.30) does not hold but the optimal policies can still be developed according to (3.23) and (3.25), then Figure 3.4 gives the scope for mutual improvements.

One may conclude that (3.30) is crucial for the possibility of an agreement on the contract curve because certain discount factor combinations make such an agreement infeasible a priori. Condition (3.30) demonstrates that both agents must show a certain degree of patience in order to be able to realize an agreement on the contract curve. Note carefully that (3.30) gives only a necessary condition, and feasibility does not mean that only these contracts on the contract curve will be agreed upon when they are supported by a weakly renegotiation-proof punishment plan.

Determining points F and C in Figures 3.3 and 3.4 has implicitly assumed that going along the compliance constraint curve increases agent i's utility and decreases agent j's utility and vice versa. Both figures indicate that the compliance constraints imply a range of possible, conflicting solutions which the agents are expected to bargain for if they had to choose among agreements on the compliance constraints. Consider, for example, the range FE in Figure 3.3. Obviously, agent j prefers the best policy, that is, F, but agent i prefers point E which gives him/her a higher utility.

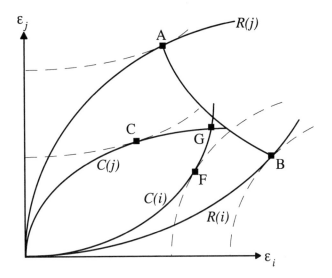

Figure 3.4   *Scope for cooperation without transfers: the case of medium discount factors*

This result does not hold in general. The condition which was implicitly assumed for developing optimal policies in Figures 3.3 and 3.4 can be developed by considering the intersection of both compliance constraints. According to (3.12), the compliance constraints intersect at an effort level of agent j which is given by

$$\varepsilon_j = \frac{1}{\delta_i} \frac{\gamma_i}{2\alpha_i} \left[\frac{1}{\delta_j} \frac{\gamma_j}{2\alpha_j} \varepsilon_j^2 \right]^2 \Leftrightarrow \varepsilon_j = 2 \left[\delta_i \delta_j^2 \frac{\alpha_i \alpha_j^2}{\gamma_i \gamma_j^2} \right]^{1/3} . \qquad (3.31)$$

Condition (3.31) specifies the maximum E_j which guarantees compliance. For (3.23) to hold for the optimal policies, E_j^F must fall short of E_j according to (3.31):

$$\frac{\alpha_j}{\gamma_j} + 2 \left[\delta_i \delta_j^2 \frac{\alpha_i \alpha_j^2}{\gamma_i \gamma_j^2} \right]^{1/3} \geq E_j^F$$

$$\Leftrightarrow 2\left[\delta_i\delta_j^2\frac{\alpha_i\alpha_j^2}{\gamma_i\gamma_j^2}\right]^{1/3} \geq \frac{1}{\delta_i}\frac{\gamma_i}{2\alpha_i}\left[\delta_i^2\frac{2\alpha_i^2\alpha_j}{\gamma_i^2\gamma_j}\right]^{2/3} \Leftrightarrow \delta_j \geq \frac{1}{4}. \qquad (3.32)$$

If and only if δ_j does not fall short of 0.25, (3.23) gives the optimal policy for agent j. Similarly, only $\delta_i \geq 0.25$ ensures that (3.25) gives the optimal policy for agent i. Figure 3.5 shows a bargaining problem for $\delta_i < 0.25$ and $\delta_j > 0.25$. The first assumption also ensures that the contract curve is completely dominated.

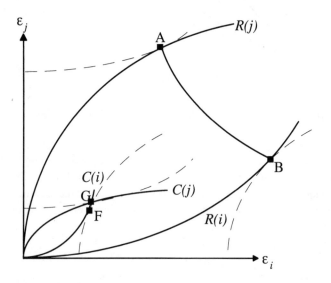

Figure 3.5   Scope for cooperation without transfers for $\delta_i < 0.25$ and $\delta_j > 0.25$

The best policy for agent i is given by point G of Figure 3.5 which is below the utility level (3.23) which a δ_i exceeding 0.25 specified. The iso-utility curve through G demonstrates that all other points on agent j's compliance constraint curve make both agents worse off. In this case, agent i's compliance constraint curve is so steep, because of his/her substantially low discount factor, that the tangential solution according to (3.23) lies rightwards of the intersection of both agents' compliance constraints (and is therefore infeasible). Figure 3.5 shows that an extremely demanding compliance constraint of agent i leaves no scope for conflicts on agent j's compliance constraint curve because both agents prefer G in Figure 3.5 if a feasible agreement has to be found on this curve.

Figure 3.5 indicates that conflicts which initiate bargaining still exist if an agreement has to be found on agent i's compliance constraint curve. However, if both discount factors fall short of 0.25, there is no conflict about the agreement as is shown in Figure 3.6.

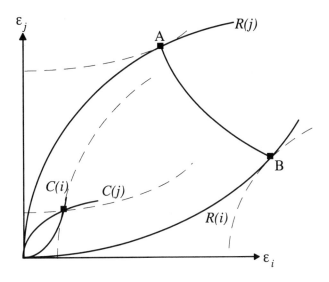

Figure 3.6   Scope for cooperation without transfers for δ_i, $\delta_j < 0.25$

The arguments given for (3.32) for agent i's optimal policies apply accordingly to the optimal policies for agent j. If both discount factors fall short of 0.25, there is no disagreement on the bargaining solution. Any other agreement than the one which specifies the intersection of both compliance constraints makes both agents worse off. As all agreements in the core embraced by both compliance constraints are Pareto dominated by an agreement which specifies the intersection of both compliance constraints as well, $\delta_i < 0.25$ and $\delta_j < 0.25$ leave only one candidate for the strategic bargaining solution. Thus, Figure 3.6 directly gives the solution for the problem because no other agreement can make any of the two agents better off. Therefore, if δ_i, $\delta_j < 0.25$, the bargaining solution is straightforward:

$$\varepsilon_j^* = 2\left[\delta_i\delta_j^2 \frac{\alpha_i\alpha_j^2}{\gamma_i\gamma_j^2}\right]^{1/3} \qquad \text{for } \delta_i, \delta_j < 0.25, \qquad (3.33)$$

$$\varepsilon_i^* = 2\left[\delta_i^2\delta_j \frac{\alpha_i^2\alpha_j}{\gamma_i^2\gamma_j}\right]^{1/3} \qquad \text{for } \delta_i, \delta_j < 0.25. \qquad (3.34)$$

The bargaining-based effort levels are denoted by a star. If δ_i $(\delta_j) > 0.25$ but δ_j $(\delta_i) < 0.25$, (3.25) [(3.23)] gives the optimal policy for agent i (agent j) which is only one candidate for the strategic bargaining solution. Equations (3.33) and (3.34) and Figure 3.6 complete developing optimal policies and give one solution for the case of extremely low discount factors. Table 3.1 summarizes the results of this section.

Table 3.1 shows which conditions the discount factors have to fulfil in order to make certain agreements feasible. Only in the case of δ_i, $\delta_j < 0.25$, does the discount factor combination directly imply the strategic bargaining solution. As low discount factors correspond with a low degree of institutional toughness, the institutional toughness implies not only the chosen contract but also determines in a first step the set of all feasible contracts. Table 3.1 demonstrates that restrictions may be severe. Note that the critical level of 0.25 is not too low a number from the standpoint of empirical relevance. When repetition takes place, say, every ten years (that is, the period of unity length covers ten years), an annual risk-adjusted interest rate of 0.15 is sufficient to make the respective δ fall short of 0.25.

3.3 CONCESSION LINE AND STRATEGIC BARGAINING SOLUTION

The last section has demonstrated that the scope for cooperation is determined by the compliance constraints which themselves depend on the agents' discount factors. This section assumes that the compliance constraints do not reduce the strategic bargaining solution to one candidate because all other solutions are Pareto inferior. Instead, it assumes that both discount factors exceed 0.25 and it will thereby refer to the scenarios described by Figures 3.3 and 3.4. The last section developed the optimal policies of both agents, the compliance constraints and the contract curve. The contract curve is the set of Pareto-efficient agreements which contains all possible solutions

Table 3.1 Contract curve, optimal policies and compliance constraints for different discount factor combinations

δ_i	δ_j	$\delta_i \delta_j$	Contract curve ...	Optimal policies are determined by ...	Conflicts on the compliance constraints ...
> 0.25	> 0.25	> 0.25	is not dominated	(3.23) and (3.25)	are possible
> 0.25	> 0.25	< 0.25	is dominated	(3.23) and (3.25)	are possible
< 0.25	> 0.25	< 0.25	is dominated	(3.23) and (3.33)	are only possible on agent j's compliance constraint
< 0.25	> 0.25	< 0.25	is dominated	(3.25) and (3.34)	are only possible on agent i's compliance constraint
< 0.25	< 0.25	< 0.25	is dominated	(3.33) and (3.34)	are not possible

when contracts are enforceable. However, as neither agent is interested in Pareto efficiency but in maximizing his/her bargaining gains subject to compliance of the other agent, the optimal policies do not belong to the contract curve when compliance constraints bind (see the points C and F in Figures 3.3 and 3.4). This section will determine the bilateral bargaining outcome by applying the strategic bargaining solution developed in Section 2.5 to the so-called concession line.

Obviously, the bargaining gains for an agent are maximized by the optimal policies. These points are not Pareto efficient in a sense that they belong to the contract curve unless both discount factors are one. Because optimal policies describe possible extremes of any strategic bargaining outcome, strategic bargaining outcomes are obviously not restricted to the contract curve. The concrete agreement which results from the bargaining process is not determined by the feasibility of agreements on the contract curve but by the relative impatience of both agents. Starting with Figure 3.3, it is evident that every self-enforcing contract will lie on the line CDEF because any contract strictly in the area below this line or on the compliance constraint curves below C and F would leave mutual improvements unexploited. If a proposal strictly in the area below CDEF were accepted, both agents could still bargain for unexploited bargaining gains. The line CDEF which is determined by (3.12), (3.19), (3.23), (3.25), (3.27) and (3.28) will be interpreted as the concession line of both agents: agent j (i) concedes by offering effort vectors starting from F (C) in the direction of C (F).

This section will make use of both the contract curve and the compliance constraints in order to determine the total derivatives of each agent's bargaining gains in terms of one agent's effort level. The first and the second total derivative of agent i's bargaining gains (see (3.1)) with respect to his/her effort level are given by

$$\frac{dU_i}{dE_i} = \alpha_i \left[1 + \frac{dE_j}{dE_i} \right] - \gamma_i E_i,$$

$$\frac{d^2 U_i}{dE_i^2} = \alpha_i \frac{d^2 E_j}{dE_i^2} - \gamma_i. \tag{3.35}$$

On the concession line, dE_j/dE_i and d^2E_j/dE_i^2 are defined by the contract curve or the compliance constraints. Although the corresponding utility functions are continuous, their first derivatives may not exist at points D and E when moving along the concession line. Because the elasticity term ρ which determines the strategic bargaining equilibrium includes the first derivatives of U_i and U_j, this section must address the behaviour of ρ on the concession line in general and in particular in points D and E.

Before turning to this discussion, the changes of individual bargaining gains along the concession line can be analysed in more detail. Point C defines the agreement which gives agent i his/her best feasible self-enforcing contract. It is evident that any change from C to F along the concession line will decrease agent i's bargaining gains which reach their feasible minimum at point F. But it is not evident whether the marginal bargaining gains increase or decrease along the concession line.

In the range CD, the compliance constraint of j defines the movement along the concession line. In order to develop dE_j/dE_i only in terms of E_i, the lower line (3.12) must be inserted into the inverse of the lower line of (3.13) and produces

$$\left(\frac{dE_j}{dE_i}\right)^{C(j)} = \delta_j \frac{\alpha_j}{\gamma_j}\left[2\delta_j \frac{\alpha_j}{\gamma_j}\varepsilon_i\right]^{-1/2},$$

$$\left(\frac{d^2E_j}{dE_i^2}\right)^{C(j)} = -\delta_j^2 \frac{\alpha_j^2}{\gamma_j^2}\left[2\delta_j \frac{\alpha_j}{\gamma_j}\varepsilon_i\right]^{-3/2} < 0.$$

According to (3.35),

$$\left(\frac{d^2U_i}{dE_i^2}\right)^{C(j)} < 0$$

holds.

In the range DE, the contract curve defines the movement along the concession line according to (3.20). In this range,

$$\left(\frac{d^2U_i}{dE_i^2}\right)^{CO} = \frac{2\gamma_i^2}{\gamma_j}\frac{\alpha_i^2\alpha_j}{\left(\gamma_iE_i-\alpha_i\right)^3} - \gamma_i$$

is undetermined in sign and depends on the parameters and on the level of agent i's efforts. A high E_i makes a negative second derivative more likely. In the range EF, the compliance constraint of i defines the movement along the concession line and produces

$$\left(\frac{d^2E_j}{dE_i^2}\right)^{C(i)} = \frac{1}{\delta_i}\frac{\gamma_i}{\alpha_i} \quad \text{and} \quad \left(\frac{d^2U_i}{dE_i^2}\right)^{C(i)} = \gamma_i\left[\frac{1}{\delta_i}-1\right] > 0.$$

Hence, agent i (agent j) faces decreasing marginal bargaining gains in range CD (range FE) and increasing marginal bargaining gains in range EF (range DC). Concessions on a part of the concession line which an agent's own compliance constraint defines lead to decreasing marginal losses because bargaining gains are decreasing and convex. The opposite holds for concessions on the other agent's compliance constraint. Turned around, an agent realizes an overproportional increase of bargaining gains through concessions of the other agent on his/her own compliance constraint curve. Conversely, he/she expects an underproportional increase of bargaining gains through concessions of the other agent on the other agent's compliance constraint curve.

Although the second derivatives influence the change of ρ along the concession line, they do not endanger the existence of a perfect equilibrium on the concession line if the set of efficient agreements defines a concave and at least piecewise differentiable frontier. The salient result of this section is the possibility that a strategic bargaining solution may lie strictly on a compliance constraint curve. Therefore, the section will start with a discussion on the development of ρ on FE.

The Development of Bargaining Gains along the Concession Line

If the contract curve is not completely dominated, the effort level of agent i lies in the range (see (3.23) and (3.28))

$$
E_i \in \left[\left[\delta_i^2 \frac{2\alpha_i^2 \alpha_j}{\gamma_i^2 \gamma_j} \right]^{1/3} + \frac{\alpha_i}{\gamma_i}, \left[\delta_i \frac{2\alpha_i^2 \alpha_j}{\gamma_i^2 \gamma_j} \right]^{1/3} + \frac{\alpha_i}{\gamma_i} \right].
$$

For the limit case $\delta_i = 1$, E and F coincide. For all $\delta_i < 1$, changes from F to E can be given by the compliance constraint (see the first line of (3.12)). Because E_i is changed continuously along the compliance constraint and E_j is a monotone function of E_i, the behaviour of E_i and E_j can be described by a single parameter d_i:

$$
E_i = \left[d_i \frac{2\alpha_i^2 \alpha_j}{\gamma_i^2 \gamma_j} \right]^{1/3} + \frac{\alpha_i}{\gamma_i}, \quad \delta_i^2 \le d_i \le \delta_i. \tag{3.36}
$$

$$
E_j = \frac{1}{\delta_i} \frac{\gamma_i}{2\alpha_i} \left[d_i \frac{2\alpha_i^2 \alpha_j}{\gamma_i^2 \gamma_j} \right]^{2/3} + \frac{\alpha_j}{\gamma_j}, \delta_i^2 \le d_i \le \delta_i. \tag{3.37}
$$

Equations (3.36) and (3.37) give E_i and E_j, respectively, as functions of d_i. The introduction of d_i allows U_i, U_j, dU_i and dU_j to be specified as functions of d_i only instead of both E_i and E_j. d_i is introduced in order to discuss the behaviour of ρ along FE. Equations (3.36) and (3.37) together with (3.1), (3.12) and (3.13) imply:

$$U_i = \left[\frac{1}{\delta_i} - 1\right]\frac{\gamma_i}{2}\left[d_i \frac{2\alpha_i^2\alpha_j}{\gamma_i^2\gamma_j}\right]^{2/3}, \qquad (3.38)$$

$$U_j = \alpha_j \left[d_i \frac{2\alpha_i^2\alpha_j}{\gamma_i^2\gamma_j}\right]^{1/3}\left[1 - \frac{d_i}{4\delta_i^2}\right], \qquad (3.39)$$

$$\frac{dU_i}{dE_i} = \left[\frac{1}{\delta_i} - 1\right]\gamma_i\left[d_i \frac{2\alpha_i^2\alpha_j}{\gamma_i^2\gamma_j}\right]^{1/3} > 0, \qquad (3.40)$$

$$\frac{dU_j}{dE_i} = \alpha_j\left[1 - \frac{d_i}{\delta_i^2}\right] \le 0. \qquad (3.41)$$

Note that U_j cannot be negative: the smallest U_j is given for $d_i = \delta_i$ which implies

$$U_j(d_i = \delta_i) = \alpha_j\left[\delta_i \frac{2\alpha_i^2\alpha_j}{\gamma_i^2\gamma_j}\right]^{1/3}\left[1 - \frac{1}{4\delta_i}\right].$$

A negative U_j would require $\delta_i < 0.25$ which is the condition for non-existence of any feasible agreement on the contract curve.

The same arguments hold for the other compliance constraint. If the contract curve is not completely dominated, E_j lies in the range

$$E_j \in \left[\left[\delta_j^2 \frac{2\alpha_j^2\alpha_i}{\gamma_j^2\gamma_i}\right]^{1/3} + \frac{\alpha_j}{\gamma_j}, \left[\delta_j \frac{2\alpha_j^2\alpha_i}{\gamma_j^2\gamma_i}\right]^{1/3} + \frac{\alpha_j}{\gamma_j}\right]$$

and can be given by

$$E_j = \left[d_j \frac{2\alpha_j^2 \alpha_i}{\gamma_j^2 \gamma_i} \right]^{1/3} + \frac{\alpha_j}{\gamma_j}, \quad \delta_j^2 \le d_j \le \delta_j, \quad (3.42)$$

which defines E_i, the bargaining gains and the marginal utilities:

$$E_i = \frac{1}{\delta_j} \frac{\gamma_j}{2\alpha_j} \left[d_j \frac{2\alpha_j^2 \alpha_i}{\gamma_j^2 \gamma_i} \right]^{2/3} + \frac{\alpha_i}{\gamma_i}, \quad \delta_j^2 \le d_j \le \delta_j, (3.43)$$

$$U_i = \alpha_i \left[d_j \frac{2\alpha_j^2 \alpha_i}{\gamma_j^2 \gamma_i} \right]^{1/3} \left[1 - \frac{d_j}{4\delta_j^2} \right], \quad (3.44)$$

$$U_j = \left[\frac{1}{\delta_j} - 1 \right] \frac{\gamma_j}{2} \left[d_j \frac{2\alpha_j^2 \alpha_i}{\gamma_j^2 \gamma_i} \right]^{2/3}, \quad (3.45)$$

$$\frac{dU_i}{dE_j} = \alpha_i \left[1 - \frac{d_j}{\delta_j^2} \right] \le 0, \quad (3.46)$$

$$\frac{dU_j}{dE_j} = \left[\frac{1}{\delta_j} - 1 \right] \gamma_j \left[d_j \frac{2\alpha_j^2 \alpha_i}{\gamma_j^2 \gamma_i} \right]^{1/3} > 0. \quad (3.47)$$

U_i is strictly positive because a negative U_i would require $\delta_j < 0.25$. As (3.44) implies $\delta_j > 0.25$ as well, the feasibility of agreements on the contract curve is not ruled out.

A similar simplification is possible for the contract curve. The effort level of agent i lies in the range

$$E_j \in \left[\frac{1}{\delta_j} \frac{\gamma_j}{2\alpha_j} \left[\delta_j \frac{2\alpha_j^2 \alpha_i}{\gamma_j^2 \gamma_i} \right]^{2/3} + \frac{\alpha_i}{\gamma_i}, \left[\delta_i \frac{2\alpha_i^2 \alpha_j}{\gamma_i^2 \gamma_j} \right]^{1/3} + \frac{\alpha_i}{\gamma_i} \right].$$

$$\frac{1}{\delta_j}\frac{\gamma_j}{2\alpha_j}\left[\delta_j\frac{2\alpha_j^2\alpha_i}{\gamma_j^2\gamma_i}\right]^{2/3}=\left[\frac{\alpha_i^2\alpha_j}{2\delta_j\gamma_i^2\gamma_j}\right]^{1/3}$$

allows E_i to be written as

$$E_i=\eta\left[\frac{\alpha_i^2\alpha_j}{\gamma_i^2\gamma_j}\right]^{1/3}+\frac{\alpha_i}{\gamma_i},\ \left(2\delta_j\right)^{-1/3}\le\eta\le\left(2\delta_i\right)^{1/3}. \tag{3.48}$$

η is defined only for the range

$$\left[\left(2\delta_j\right)^{-1/3},\left(2\delta_i\right)^{1/3}\right].$$

This range is not empty if $\delta_i\delta_j\ge 0.25$ which is the condition for the feasibility of an agreement on the contract curve. Accordingly, E_j can be written as

$$E_j=\frac{1}{\eta}\left[\frac{\alpha_i\alpha_j^2}{\gamma_i\gamma_j^2}\right]^{1/3}+\frac{\alpha_j}{\gamma_j}. \tag{3.49}$$

According to these definitions, the bargaining gains and the marginal utilities are given by

$$U_i=\left[\frac{\alpha_i^2\alpha_j}{\gamma_i^2\gamma_j}\right]^{1/3}\gamma_i\left[\frac{1}{\eta}-\frac{\eta^2}{2}\right], \tag{3.50}$$

$$U_j=\left[\frac{\alpha_i^2\alpha_j}{\gamma_i^2\gamma_j}\right]^{1/3}\alpha_j\left[\eta-\frac{1}{2\eta^2}\right]. \tag{3.51}$$

$$\frac{dU_i}{d\eta} = -\left[\frac{\alpha_i^2 \alpha_j}{\gamma_i^2 \gamma_j}\right]^{1/3} \gamma_i \left[\eta + \frac{1}{\eta^2}\right] < 0, \tag{3.52}$$

$$\frac{dU_j}{d\eta} = \left[\frac{\alpha_i^2 \alpha_j}{\gamma_i^2 \gamma_j}\right]^{1/3} \alpha_j \left[1 + \frac{1}{\eta^3}\right] > 0. \tag{3.53}$$

Note that none of the calculations for the contract curve do include any agent's discount factor. Additionally, the derivatives (3.52) and (3.53) are given in terms of η instead of E_i or E_j. As only dU_i and dU_j determine the strategic bargaining outcome (and both $d\eta$ are cancelled when developing ρ), the easiest definitions in terms of η may be used. Coming back to the arc FE, (3.40) and (3.41) define the elasticity term ρ as a function of d_i:

$$\rho(d_i) = \frac{\left[\frac{1}{\delta_i} - 1\right]\gamma_i \left[d_i \dfrac{2\alpha_i^2 \alpha_j}{\gamma_i^2 \gamma_j}\right]^{1/3}}{\alpha_j \left[1 - \dfrac{d_i}{\delta_i^2}\right]} \frac{V_j(\delta_i, d_i)}{V_i(\delta_i, d_i)}. \tag{3.54}$$

Equation (3.54) gives the utilities as functions of δ_i and d_i. A first significant conclusion can be drawn: because both V_i and V_j are strictly positive for all feasible agreements, the elasticity ρ approaches $-\infty$ when the contract on the concession line approaches F. Point F equalizes d_i and δ_i^2 and makes the denominator of the first term in (3.54) zero. As similar arguments hold for the arc CD, point C equalizes the corresponding d_j and δ_j^2 and makes the numerator zero. Hence, the elasticity ρ at point C is zero:

$$\lim_{\substack{E_i \to E_i^F \\ E_j \to E_j^F}} \rho\left[E_i, E_j\right] = -\infty, \quad \rho\left[E_i^C, E_j^C\right] = 0. \tag{3.55}$$

In addition to the behaviour of ρ on the concession line, these conclusions will be fundamental for the model.

The strategic bargaining solution can be given as an implicit function:

$$\ln \delta_j \, \frac{dV_i(d_i, \cdot)}{V_i(d_i, \cdot)} + \ln \delta_i \, \frac{dV_j(d_i, \cdot)}{V_j(d_i, \cdot)} = 0. \tag{3.56}$$

Equation (3.56) includes the strategic bargaining condition. Equations (3.55) and (3.56) show a result which holds generally: if $\delta_j \to 1$, $\delta_i < 1$, point F is the strategic bargaining equilibrium. In this case, $-\ln\delta_i/\ln\delta_j \to -\infty$. Hence, one finds that perfect patience of agent j is a necessary and sufficient condition for the realization of point F, that is, the optimal policy for agent j (if $\delta_i < 1$). As agent j concedes from F to E, it is obvious that discount factor combinations with $\delta_i < 1$ exist which specify an agreement between F and E (the complete proof of this result will be given soon). Hence, strategic bargaining between two agents who differ by their discount factors may imply a solution which lies on the compliance constraint curve of an agent although agreements on the contract curve are feasible. This result, which is seemingly Pareto inferior in the absence of compliance constraints, may represent a strategic bargaining solution which results from bargaining gain maximization subject to potential delay costs and compliance of both agents.

Since the same arguments hold for the compliance constraint that d_j describes, $\delta_i \to 1$, $\delta_j < 1$ implies point C (in this case, $-\ln\delta_j/\ln\delta_i$ is zero). Perfect patience of agent i is therefore a necessary and sufficient condition to realize the optimal policy for agent i. Note that both the case $\delta_i \to 1$ and the case $\delta_j \to 1$ assume that the other agent's discount factor falls short of one. If they did not (that is, $\delta_i \to 1$, $\delta_j \to 1$), the solution has to be found on the contract curve since in this case the compliance constraints fall together with the utility reservation curves.

An expression similar to (3.56) can be derived for strategic bargaining solutions on the contract curve:

$$\ln \delta_j \, \frac{dV_i(\eta, \cdot)}{V_i(\eta, \cdot)} + \ln \delta_i \, \frac{dV_j(\eta, \cdot)}{V_j(\eta, \cdot)} = 0. \tag{3.57}$$

Equation (3.57) is only defined for the range given in (3.48). Note that (3.56) and (3.57) can be integrated in order to determine the strategic bargaining solution as long as bargaining gains can be given as functions of a single variable.

The Behaviour of ρ along the Concession Line

Until now, this section has merely demonstrated that ρ approaches $-\infty$ at point F and is equal to zero at point C. It has not yet demonstrated how ρ is changed when it moves along the concession line from F to C. In order to guarantee uniqueness of the strategic bargaining solution, it is necessary that every ρ which could be equal to $-\ln\delta_i/\ln\delta_j$ corresponds to one and only one effort combination on the concession line.

Therefore, it has to be explored in a first step how ρ develops on the compliance constraints and the contract curve. In a second step, it has to be explored how ρ behaves when switching from a compliance constraint to the contract curve and vice versa. Since the arguments for the arc CD are qualitatively the same as for FE, it is sufficient to consider the behaviour of ρ along FE, along ED and at point E. Equation (3.58) shows that the change of ρ along FE and ED is unambiguous in sign.

$$\frac{d\rho(d_i)}{dd_i} = \frac{d\left(\frac{dU_i}{dU_j}(\cdot)\right)}{dd_i}\frac{V_j(\cdot)}{V_i(\cdot)} + \frac{\frac{dV_j(\cdot)}{dd_i}V_i(\cdot) - \frac{dV_i(\cdot)}{dd_i}V_j(\cdot)}{V_i(\cdot)^2}\frac{dU_i}{dU_j}(\cdot) > 0,$$

$$\frac{d\rho(\eta)}{d\eta} = \frac{d\left(\frac{dU_i}{dU_j}(\cdot)\right)}{d\eta}\frac{V_j(\cdot)}{V_i(\cdot)} + \frac{\frac{dV_j(\cdot)}{d\eta}V_i(\cdot) - \frac{dV_i(\cdot)}{d\eta}V_j(\cdot)}{V_i(\cdot)^2}\frac{dU_i}{dU_j}(\cdot) < 0.$$

$$(3.58)$$

The sign of $d\rho(d_i)/dd_i$ follows from $dV_j/dd_i < 0$, $dV_i/dd_i > 0$ (see (3.38) and (3.39)) which make the second term positive (note that dU_i/dU_j is negative). The sign of the first term depends on

$$\frac{dU_i}{dU_j}(\cdot) = \left[\frac{1}{\delta_i} - 1\right]\frac{\gamma_i}{\alpha_j}\left[\frac{d_i\frac{2\alpha_i^2\alpha_j}{\gamma_i^2\gamma_j}}{1 - \frac{d_i}{\delta_i^2}}\right]^{1/3}, \quad \Gamma(d_i) := d_i^{1/3}\left[1 - \frac{d_i}{\delta_i^2}\right]^{-1}$$

(see (3.40) and (3.41)). Since $d\Gamma(d_i)/dd_i > 0$, the first term in $d\rho(d_i)/dd_i$ is positive as well and $d\rho(d_i)/dd_i > 0$ holds.

The sign of $d\rho(\eta)/d\eta$ follows from $dV_j/d\eta > 0$, $dV_i/d\eta < 0$ (see (3.52) and (3.53)) which make the second term negative. The sign of the first term depends on

$$\frac{dU_i}{dU_j}(\cdot) = -\gamma\eta$$

(see (3.52) and (3.53)) whose derivative with respect to η is negative. Hence the first term in $d\rho(\eta)/d\eta$ is negative as well and $d\rho(\eta)/d\eta < 0$ holds.

When introducing d_i and η, it has been shown that agent i concedes from F by increasing d_i until E is reached after which he/she decreases η when moving in the direction of D. Hence, ρ is increased continuously when moving from F to E and when moving from E to D. Uniqueness is guaranteed if no ρ exists which belongs to both ranges. This condition can be shown by discussing the behaviour of ρ at E.

At point E, two derivatives exist which signal a potential jump of the elasticity. The utilities of both agents are given by the iso-utility curves which are now tangential in point E because E belongs to the contract curve. Differentiation involves discrimination of two cases: whether the differential is determined by the compliance constraint curve or by the contract curve. A marginal change 'from the direction' of X to Y at point Y will be denoted by the subscript XY. Thus, *FE* denotes the differentiation which uses the compliance constraint (coming from point F), *DE* denotes the differentiation which uses the contract curve (coming from point D). The differentials are (see (3.12), (3.28) and (3.20))

$$\left(\frac{dE_j}{dE_i}\right)_{FE} = \frac{1}{\delta_i}\frac{\gamma_i}{\alpha_i}\left[E_i^E - \frac{\alpha_i}{\gamma_i}\right] = \frac{1}{\delta_i}\frac{\gamma_i}{\alpha_i}\left[\delta_i\frac{2\alpha_i^2\alpha_j}{\gamma_i^2\gamma_j}\right]^{1/3},$$

$$\left(\frac{dE_j}{dE_i}\right)_{DE} = -\frac{\gamma_i}{\gamma_j}\frac{\alpha_i\alpha_j}{\left[\gamma_iE_i^E - \alpha_i\right]^2} = -2\delta_i\left[\frac{\alpha_j\gamma_i}{2\delta_i\alpha_i\gamma_j}\right]^{1/3}. \qquad (3.59)$$

Equations (3.59) can serve to determine the marginal utilities which must also be discriminated:

$$\left(\frac{dU_i}{dE_i}\right)_{FE} = \left[\frac{1}{\delta_i} - 1\right]\gamma_i\left[\delta_i\frac{2\alpha_i^2\alpha_j}{\gamma_i^2\gamma_j}\right]^{1/3}, \quad \left(\frac{dU_j}{dE_i}\right)_{FE} = \alpha_j\left[1 - \frac{1}{\delta_i}\right], \qquad (3.60)$$

$$\left(\frac{dU_i}{dE_i}\right)_{DE} = -\gamma_i \left[\delta_i \frac{2\alpha_i^2 \alpha_j}{\gamma_i^2 \gamma_j}\right]\left[1+\left(2\delta_i\right)^{1/3}\right],$$

$$\left(\frac{dU_j}{dE_i}\right)_{DE} = \alpha_j \left[1+\left(2\delta_i\right)^{1/3}\right]. \tag{3.61}$$

That the signs of the derivatives differ between FE and DE should not be surprising: on the arc FE, agent i (agent j) gains (loses) by an increase of i's effort level, on the arc DE, agent i (agent j) loses (gains) by an increase of i's effort level (see Figure 3.3). Using (3.60) and (3.61) to determine the elasticity at point E reveals that

$$\rho_{FE} = \rho_{DE} = -\frac{\gamma_i}{\alpha_j}\left[\delta_i \frac{2\alpha_i^2 \alpha_j}{\gamma_i^2 \gamma_j}\right]^{1/3}\frac{V_j}{V_i} \tag{3.62}$$

holds. A similar result is valid for point D, too. Despite the differing derivatives of the bargaining gain functions, the elasticity coincides at point E when the compliance constraint approaches the contract curve and vice versa. Additionally, it is known that $-\ln\delta_i/\ln\delta_j$ covers the range $[-\infty,0]$ for all conceivable discount factor combinations including perfect patience of one agent. As it is also known that ρ does not jump on both the contract curve and the compliance constraint curves, (3.55), (3.58) and (3.62) prove that every outcome on the concession line is principally possible and that the whole range of elasticities, that is $-\infty \le \rho \le 0$ which could be implied by $-\ln\delta_i/\ln\delta_j$ is indeed covered by a point on the concession curve CDEF. If the contract curve is not completely dominated by the compliance constraints, the mapping from $\ln\delta_i/\ln\delta_j$ to ρ and vice versa is unique: any specific $\ln\delta_i/\ln\delta_j$ corresponds to a unique ρ and any specific ρ corresponds to a unique $\ln\delta_i/\ln\delta_j$. Figure 3.7 shows the behaviour of ρ along CDEF.

Note that Figure 3.7 does not give a function because the movement from C to F alters both E_i and E_j. Hence, Figure 3.7 is only an illustration of how ρ behaves along the concession line. It demonstrates uniqueness because every possible $-\ln\delta_i/\ln\delta_j$ corresponds to one and only one ρ. In particular, Figure 3.7 shows that a large set of discount factor combinations imply a solution on the compliance constraints although agreements on the contract curve are possible.

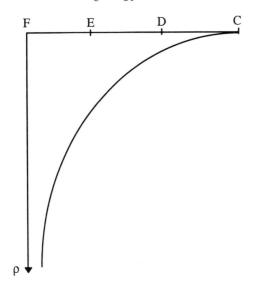

Figure 3.7   Behaviour of ρ along CDEF

The condition for the existence of feasible agreements on the contract curve was determined by (3.30). If (3.30) is violated, the contract curve is completely dominated by the compliance constraints. If $\delta_i\delta_j < 0.25$ and δ_i, $\delta_j > 0.25$ hold, Table 3.1 has indicated that there is some scope for bargaining on both compliance constraints. This situation was shown in Figure 3.4 in which point G indicates the intersection of both compliance constraints. Obviously, point G defines the maximum sum of individual efforts which is feasible. This point would solve a constrained joint maximization problem which allows for enforceable transfers. (Chapter 4 will demonstrate that point G is not necessarily the bargaining solution if transfers are possible but transfer policies must be self-enforcing as well.) In a non-cooperative environment, however, point G must qualify for the strategic bargaining solution by fulfilling condition (2.52).

The absence of feasible agreements on the contract curve defines the new concession line CGF (see Figure 3.4). For this concession line, (3.58) still holds so that the change of ρ along FG and GC is unambiguous in sign. Hence, the behaviour of ρ along CGF depends on the behaviour of ρ at point G. The intersection of compliance constraints has already been developed by (3.33) and (3.34) so that

$$E_i^G = 2\left[\delta_j \delta_i^2 \frac{\alpha_i^2 \alpha_j}{\gamma_i^2 \gamma_j}\right]^{1/3} + \frac{\alpha_i}{\gamma_i}, \quad E_j^G = 2\left[\delta_i \delta_j^2 \frac{\alpha_i \alpha_j^2}{\gamma_i \gamma_j^2}\right]^{1/3} + \frac{\alpha_j}{\gamma_j}. \quad (3.63)$$

gives the effort levels for point G. Differentiations at point G involve two cases, too. *FG* will denote the marginal change at point G which agent *i*'s compliance constraint induces, and *CG* will denote the marginal change at point G which agent *j*'s compliance constraint induces:

$$\left.\begin{array}{l}\left(\dfrac{dE_j}{dE_i}\right)_{FG} = \dfrac{2}{\delta_i}\dfrac{\gamma_i}{\alpha_i}\left[\delta_j \delta_i^2 \dfrac{\alpha_i^2 \alpha_j}{\gamma_i^2 \gamma_j}\right]^{1/3}\\[20pt]\left(\dfrac{dE_j}{dE_i}\right)_{CG} = \delta_j \dfrac{\alpha_j}{2\gamma_j}\left[\delta_i \delta_j^2 \dfrac{\alpha_i \alpha_j^2}{\gamma_i \gamma_j^2}\right]^{-1/3}\end{array}\right\} \Rightarrow \qquad (3.64)$$

$$\left(\frac{dU_i}{dE_i}\right)_{FG} = 2\gamma_i\left[\frac{1}{\delta_i}-1\right]\left[\delta_j \delta_i^2 \frac{\alpha_i^2 \alpha_j}{\gamma_i^2 \gamma_j}\right]^{1/3},$$

$$\left(\frac{dU_j}{dE_i}\right)_{FG} = \alpha_j\left[1-4\delta_j\right], \qquad (3.65)$$

$$\left(\frac{dU_i}{dE_i}\right)_{CG} = 2\gamma_i\left[\frac{1}{4\delta_i}-1\right]\left[\delta_j \delta_i^2 \frac{\alpha_i^2 \alpha_j}{\gamma_i^2 \gamma_j}\right]^{1/3},$$

$$\left(\frac{dU_j}{dE_i}\right)_{CG} = \alpha_j\left[1-\delta_j\right]. \qquad (3.66)$$

(3.64) results from (3.13) computed at point G. Using (3.65) and (3.66) to derive the elasticity at point G shows that both elasticities do not coincide:

$$\rho_{FG} = \frac{2\gamma_i}{\alpha_j} \frac{1/\delta_i - 1}{1 - 4\delta_j} \left[\delta_j \delta_i^2 \frac{\alpha_i^2 \alpha_j}{\gamma_i^2 \gamma_j} \right]^{1/3} \frac{V_j\left(E_i^G, E_j^G\right)}{V_i\left(E_i^G, E_j^G\right)} := \frac{1/\delta_i - 1}{1 - 4\delta_j} \upsilon,$$

$$\rho_{CG} = \frac{2\gamma_i}{\alpha_j} \frac{1/4\delta_i - 1}{1 - \delta_j} \left[\delta_j \delta_i^2 \frac{\alpha_i^2 \alpha_j}{\gamma_i^2 \gamma_j} \right]^{1/3} \frac{V_j\left(E_i^G, E_j^G\right)}{V_i\left(E_i^G, E_j^G\right)} := \frac{1/4\delta_i - 1}{1 - \delta_j} \upsilon. \quad (3.67)$$

υ denotes the relative utility term which is identical in both elasticity terms. ρ jumps at point G if

$$\rho_{FG} < \rho_{CG}. \quad (3.68)$$

Since $\delta_j > 0.25$ for the case described by Figure 3.4, (3.68) implies

$$\frac{1/\delta_i - 1}{1 - 4\delta_j} < \frac{1/4\delta_i - 1}{1 - \delta_j} \Leftrightarrow \left[1/\delta_i - 1\right]\left[1 - \delta_j\right] > \left[1/4\delta_i - 1\right]\left[1 - 4\delta_j\right]$$

$$\Leftrightarrow \delta_i \delta_j < \frac{1}{4}. \quad (3.69)$$

$\delta_i \delta_j < 1/4$ is the definition of the case described by Figure 3.4. Hence, (3.69) proves that ρ jumps at point G when both agents move along the concession line from F to C. Conditions (3.55), (3.58) and (3.69) prove that every outcome on the concession line is possible, but that only the mapping from $\ln\delta_i/\ln\delta_j$ to ρ is unique because every specific $\ln\delta_i/\ln\delta_j$ corresponds to a unique ρ but not vice versa. Figure 3.8 shows the behaviour of ρ along CGF (and the same reservation applies to Figure 3.8 as applies to Figure 3.7). If $-\ln\delta_i/\ln\delta_j$ lies in the range of the jump, G gives the strategic bargaining solution.

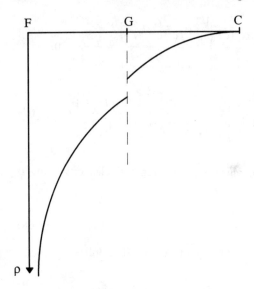

Figure 3.8   Behaviour of ρ along CGF

Developing the strategic bargaining solution can make use of d_i and d_j as well but assuming $\delta_i \delta_j < 0.25$, δ_i, $\delta_j > 0.25$ requires the ranges of d_i and d_j to be redefined. Equalizing both compliance constraints under the use of d_i and d_j gives two equations which define the maximum d_i and the maximum d_j:

$$d_i^G = 4\delta_i^2 \delta_j, \quad d_j^G = 4\delta_j^2 \delta_i. \tag{3.70}$$

The superscript G denotes the intersection of both compliance constraints as it was indicated by point G in Figure 3.4. Equation (3.70) shows that d_i^G and d_j^G fall short of δ_i and δ_j, respectively, which define the upper limits if the contract curve is not completely dominated. For $\delta_i \delta_j < 0.25$, $4\delta_i \delta_j$ falls short of one. Hence (3.36) and (3.37) must be replaced by

$$E_i = \left[d_i \frac{2\alpha_i^2 \alpha_j}{\gamma_i^2 \gamma_j} \right]^{1/3} + \frac{\alpha_i}{\gamma_i}, \quad \delta_i^2 \le d_i \le 4\delta_i^2 \delta_j, \tag{3.71}$$

$$E_j = \frac{1}{\delta_i} \frac{\gamma_i}{2\alpha_i} \left[d_i \frac{2\alpha_i^2 \alpha_j}{\gamma_i^2 \gamma_j} \right]^{2/3} + \frac{\alpha_j}{\gamma_j}, \quad \delta_i^2 \le d_i \le 4\delta_i^2 \delta_j. \tag{3.72}$$

Equations (3.38) and (3.41) must employ the upper limit $4\delta_j^2\delta_i$ as well (which is not done explicitly here). Condition (3.55) for the optimal policy of an agent who is perfectly patient still holds. Equation (3.36), however, is only defined for discount factor combinations which induce a strategic bargaining solution with $d_i \leq 4\delta_j^2\delta_i$.

This section has demonstrated that strategic bargaining solutions of self-enforcing contracts are not restricted to the contract curve. On the contrary, even if feasibility of agreements on the contract curve is guaranteed, self-enforcing contracts may employ an effort policy which is determined by the compliance constraint of an agent. As the agents are interested in maximizing their individual utility, self-enforcing contracts on the contract curve are not necessarily implied unless they are the result of strategic bargaining on the concession line. Therefore, focusing explicitly on agreements on the contract curve is likely to overlook an important set of outcomes which are chosen deliberately by both agents as a result of rational bilateral bargaining.

3.4 THE IMPACT OF DISCOUNT FACTOR CHANGES

The previous section has demonstrated that compliance constraints belong to the concession line of every strategic bargaining process when there is some scope for bargaining because the optimal policies for both agents differ. The compliance constraints themselves depend on the discount factors. The lower the discount factor of an agent, the more demanding is his/her compliance constraint. The discount factors also determine the strategic bargaining term $\ln\delta_i/\ln\delta_j$. This section will explore the impact of an agent's discount factor on the outcome of the self-enforcing contract because the impact of one agent's discount factor is obviously twofold for strategic bargaining solutions on this agent's compliance constraint curve.

The general impact of an agent's discount factor will be demonstrated by considering marginal discount factor changes. Considering discount factor changes does not mean that an agent deliberately pretends that there is a certain discount factor change in order to increase his/her individual bargaining gains. The assumption of perfect information still holds. Changing an agent's discount factor is motivated by the question of how the strategic bargaining solution and the utilities are changed if one agent – say j – has to bargain with another agent – say i – when agent i is more patient or impatient compared to the original situation. Thereby, this section can prepare a possible extension when several agents are involved but only two agents are allowed to cooperate. Restricting cooperation to two agents may be the assumed limitation of a team selection problem or the result of a coalition formation problem which is governed by different utilities which are defined by the options to join a coalition or to stay away from the self-enforcing contract (for such a model which employs a simple bargaining rule, see Stähler, 1996a).

Marginal changes assume that the original solution was not at the intersection of two curves in order to guarantee a unique derivative along a compliance constraint curve or along the contract curve. This section will demonstrate that the more impatient agent is not necessarily the loser compared to the more patient agent if a self-enforcing contract employs an effort policy on the relatively impatient agent's compliance constraint curve. This result contrasts with the standard result of the role of relative impatience when two agents have to split bargaining gains unanimously by an enforceable contract.

If $\delta_i\delta_j > 0.25$, the optimal policy for agent j (agent i) is given by specifying E_i and E_j according to (3.36) and (3.37) (according to (3.42) and (3.43)) with $d_i = \delta_i^2$ ($d_j = \delta_j^2$). Pursuing the optimal policy of an agent must be based on the assumption that this agent is perfectly patient, that is, that his/her discount factor is equal to one. Consider, for example, agent j and assume that $\delta_j = 1$ and $\delta_i < 1$ hold. According to the strategic bargaining solution, agent j is able to reap the maximal bargaining gains. Equations (3.73) and (3.74) give the bargaining gains of both agents as

$$U_i\left(d_i = \delta_i^2, \cdot\right) = \left[\frac{1}{\delta_i} - 1\right]\frac{\gamma_i}{2}\left[\delta_i^2 \frac{2\alpha_i^2\alpha_j}{\gamma_i^2\gamma_j}\right]^{2/3}, \qquad (3.73)$$

$$U_j\left(d_i = \delta_i^2, \cdot\right) = \frac{3}{4}\alpha_j\left[\delta_i^2 \frac{2\alpha_i^2\alpha_j}{\gamma_i^2\gamma_j}\right]^{1/3}. \qquad (3.74)$$

As any agreement which is a candidate for a self-enforcing contract must meet the compliance constraints, (3.73) shows that a self-enforcing contract which specifies the best policy for agent j leaves some gains to agent i in order to avoid defection of agent i. Realizing the maximal bargaining gains, therefore, does not mean reaping all bargaining gains as it would be obvious for enforceable contracts which can decouple the allocation of efforts and the distribution of bargaining gains.

What is the impact of a marginal change of δ_i on U_i and U_j? Differentiation of (3.73) and (3.74) with respect to δ_i gives

$$\frac{\partial U_i}{\partial \delta_i}\left(d_i = \delta_i^2, \cdot\right) = \frac{1}{3\delta_i^2}[1 - 4\delta_i]\frac{\gamma_i}{2}\left[\delta_i^2 \frac{2\alpha_i^2 \alpha_j}{\gamma_i^2 \gamma_j}\right]^{2/3} < 0, \quad (3.75)$$

$$\frac{\partial U_j}{\partial \delta_i}\left(d_i = \delta_i^2, \cdot\right) = \frac{1}{2\delta_i}\alpha_j\left[\delta_i^2 \frac{2\alpha_i^2 \alpha_j}{\gamma_i^2 \gamma_j}\right]^{1/3} > 0. \quad (3.76)$$

Equations (3.75) and (3.76) show that agent i improves on his/her utility by showing a marginally lower discount factor whereas the utility of agent j increases with an increase in agent i's discount factor. This effect can be made clear by considering the impact of marginal changes of δ_i on point F in Figures 3.1 or 3.2. An increase of δ_i means turning the compliance constraint defined by δ_i downwards. Obviously, a downward turn makes this constraint less demanding because agent j's efforts can be lowered marginally without endangering compliance. If agent j is perfectly patient, increasing δ_i brings him/her on to a higher iso-utility curve whereas agent i loses because a constraint is weakened for the maximization of agent j's utility.

This result does not hold for δ_i, $\delta_j < 0.25$. (Note that the negative sign of (3.75) was due to $\delta_i > 0.25$.) In this case, both optimal policies coincide since both agents prefer the intersection of their compliance constraint curves compared to all other feasible agreements (see Figure 3.6). As both agents prefer an increase of E_i and E_j unless δ_i and δ_j exceed 0.25, more scope for cooperation benefits both agents. Shifting the intersection point up- and rightwards brings both agents on to higher iso-utility curves. Thus, an increase in δ_i or δ_j benefits both agents if both agents' optimal policies coincide because both compliance constraints are extremely demanding.

The general impact of discount factor changes can be discussed by considering marginal changes of δ_i on agent i's compliance constraint curve. If the relevant part of the concession line is not changed by the discount factor under consideration, bargaining gains are increased (decreased) for an agent by an increasing (decreasing) patience of this agent. This is the standard result of strategic bargaining models and was demonstrated at the beginning of Section 2.5.

On the compliance constraint curve of an agent, however, discount factor changes may have a twofold implication. This twofold implication is given if there is a change in the discount factor of an agent who defines the compliance constraint which is relevant for the bargaining solution. Consider, for example, a marginal decrease of δ_i and assume that strategic bargaining has specified an agreement on agent i's compliance constraint. Two effects imply opposite effects on the agents' utilities:

- First, a lower δ_i diminishes the bargaining power of agent i and increases the bargaining power of agent j (bargaining effect).
- Second, a lower δ_i turns the compliance constraint curve upwards and increases the minimum efforts of agent j necessary to ensure the compliance of agent i (compliance effect).

The bargaining effect decreases the utility of agent i and increases the utility of agent j; the compliance effect increases the utility of agent i and decreases the utility of agent j. In order to compute the total effect, the signs of the partial derivatives of (3.38) and (3.39) with respect to δ_i and d_i must be determined:

$$\frac{\partial V_i(\cdot)}{\partial \delta_i} = -\frac{1}{\delta_i^2}\frac{\gamma_i}{2}\left[d_i\frac{2\alpha_i^2\alpha_j}{\gamma_i^2\gamma_j}\right]^{2/3} < 0,$$

$$\frac{\partial V_i(\cdot)}{\partial d_i} = \left[\frac{1}{\delta_i}-1\right]\frac{\gamma_i}{3}\frac{1}{d_i}\left[d_i\frac{2\alpha_i^2\alpha_j}{\gamma_i^2\gamma_j}\right]^{2/3} > 0, \qquad (3.77)$$

$$\frac{\partial V_j(\cdot)}{\partial \delta_i} = \frac{d_i}{2\delta_i^3}\alpha_j\left[d_i\frac{2\alpha_i^2\alpha_j}{\gamma_i^2\gamma_j}\right]^{1/3} > 0,$$

$$\frac{\partial V_j(\cdot)}{\partial d_i} = \frac{1}{3}\left[\frac{1}{d_i}-\frac{1}{\delta_i^2}\right]\alpha_j\left[d_i\frac{2\alpha_i^2\alpha_j}{\gamma_i^2\gamma_j}\right]^{1/3} < 0. \qquad (3.78)$$

Additionally, the total derivatives of both utility functions require the derivative of d_i to be computed with respect to δ_i. Integration of (3.56) over d_i yields

$$F_{d_i}(\delta_i, d_i) :=$$
$$\ln\delta_j\left\{\ln V_i[d_i]-\ln V_i\left[\delta_i^2\right]\right\}+\ln\delta_i\left\{\ln V_j[d_i]-\ln V_j\left[\delta_i^2\right]\right\} = 0. \quad (3.79)$$

Differentiating F_{d_i} gives

$$\frac{\partial F_{d_i}}{\partial \delta_i} = \ln \delta_j \left\{ \frac{\dfrac{\partial V_i[d_i]}{\partial \delta_i}}{V_i[d_i]} - \frac{\dfrac{\partial V_i[\delta_i^2]}{\partial \delta_i}}{V_i[\delta_i^2]} \right\} + \ln \delta_i \left\{ \frac{\dfrac{\partial V_j[d_i]}{\partial \delta_i}}{V_j[d_i]} - \frac{\dfrac{\partial V_j[\delta_i^2]}{\partial \delta_i}}{V_j[\delta_i^2]} \right\}$$

$$+ \frac{\ln V_j[d_i] - \ln V_j[\delta_i^2]}{\delta_i}$$

$$\frac{\partial F_{d_i}}{\partial d_i} = \ln \delta_j \frac{\dfrac{\partial V_i[d_i]}{\partial d_i}}{V_i[d_i]} + \ln \delta_i \frac{\dfrac{dV_j[d_i]}{dd_i}}{V_j[d_i]}, \tag{3.81}$$

both ambiguous in sign, and

$$\Rightarrow \frac{dd_i}{d\delta_i}, \frac{dV_i[d_i]}{d\delta_i} = \frac{\partial V_i[d_i]}{\partial \delta_i} + \frac{\partial V_i[d_i]}{\partial d_i} \frac{dd_i}{d\delta_i},$$

$$\frac{dV_j[d_i]}{d\delta_i} = \frac{\partial V_j[d_i]}{\partial \delta_i} + \frac{\partial V_j[d_i]}{\partial d_i} \frac{dd_i}{d\delta_i}$$

all ambiguous in sign.

Since some terms are multiplied by $\ln\delta$ terms and one term is not, it is obvious that an explicit computation does not resolve ambiguity. Equation (3.80) uses (3.75) and (3.76) and shows the ambiguous effect of a variation of agent i's discount factor. If the bargaining solution of a self-enforcing contract lies on the agent's compliance constraint curve of an agent whose discount factor is changed, the utility of both agents may increase or decrease. Note that an increase in an agent's utility does not necessarily imply a decrease in the other agent's utility because a change of a discount factor alters the concession line (possibly in favour or to the disadvantage of both agents) and should not be confused with comparing points on a given concession line.

If agent j's discount factor is changed and the solution lies on agent i's compliance constraint, the result is unambiguous because the compliance constraint defined by δ_i is untouched by a variation of δ_j. Taken together, the impact of discount factor changes depends on the range of the concession line under consideration:

- On the contract curve, neither compliance constraints bind and only the bargaining effect applies and increases the bargaining gains of the agent whose relative patience is increased.
- On the compliance constraint curve, two cases have to be distinguished: if the discount factor which defines the compliance constraint is changed, the result is ambiguous and the bargaining gains of each agent may increase or decrease. If the discount factor which does not define the compliance constraint is changed, only the bargaining effect applies and increases the bargaining gains of the agent whose relative patience is increased.

An example can demonstrate that the compliance effect may dominate the bargaining effect over a wide range. Assume that $\alpha_i = \alpha_j = \gamma_i = \gamma_j = 1$ and $\delta_j = 0.8$. Figure 3.9 gives the utilities for agent i and agent j when the discount factor of agent i is changed from 0.3 to 0.69.

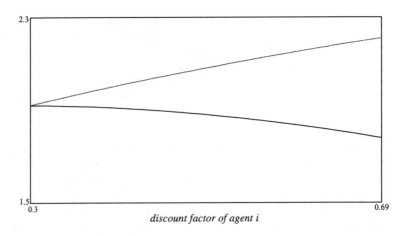

Figure 3.9   *Payoffs of agent j (upper curve) and of agent i (lower curve) for $\alpha_i = \alpha_j = \gamma_i = \gamma_j = 1$, $\delta_j = 0.8$ and $\delta_i \in [0.3, 0.69]$*

When $\delta_i \in [0.3, 0.69]$, the strategic bargaining solution is to be found on agent i's compliance constraint curve. Figure 3.9 shows that an increase in agent i's discount factor decreases his/her utility and increases agent j's utility. Hence, Figure 3.9 demonstrates that the compliance effect dominates the bargaining effect in the whole range where the compliance effect can be at work. Although agent i's bargaining power is increased by an increasing

δ_i, agent j gains because he/she has to sacrifice less utility in order to guarantee the compliance of agent i.

The discussion of discount factor variations has shown that the impact of discount factors on strategic bargaining outcomes is not that clear as it is known from strategic bargaining models which explain enforceable contracts. Referring to the problem of team selection, it is thus not quite clear with which agent another agent would like to form a partnership. In conventional models, an agent would like to split a pie of size one with the most impatient agent because this partner's impatience gives him/her the largest share of the pie. Preference for partners can be quite different if contracts must be self-enforcing.

3.5 APPLICATION I: IMPLICIT COLLUSION IN DUOPOLISTIC MARKETS

The previous sections employed a model of orthogonal reaction functions. The most prominent application of strategic bargaining for self-enforcing contracts without transfers is implicit collusion in an industrial market. Typically for duopolies, the reaction function of a duopolist has a non-zero slope. In markets for strategic substitutes, the best reaction of duopolist j is to increase his/her output if duopolist i decreases his/her output. In this case, a defecting duopolist increases his/her output level beyond the non-cooperative level. In markets for strategic complements, the best reaction of duopolist j is to decrease his/her price if duopolist i decreases his/her price. In that case, a defecting duopolist decreases his/her price below the non-cooperative level. Potential collusion in an industrial market has been a main theme in the industrial organization literature. Every textbook contains a section which deals with potential cooperation among suppliers in a non-cooperative environment. However, this discussion is almost exclusively concerned with the sustainability of the monopolistic outcome. As every firm can be expected to maximize its individual profits (and not the industry's profit), the strategic bargaining solution may also specify an agreement on a compliance constraint and not on the contract curve. It is therefore misleading to restrict one's attention exclusively to the monopolistic outcome. This section will discuss implicit collusion in both a market for strategic substitutes and a market for strategic complements.

Definition (2.12) gives the profits in a market of strategic substitutes and in a market of strategic complements. Duopolists compete by strategic substitutes (complements) if each firm's best reaction to an increase in the other firm's strategy variable is to decrease (increase) its own strategy variable (Bulow, Geanakoplos and Klemperer, 1985). The slope of the reaction curve is

$$\frac{ds'_k}{ds_{-k}} = -\frac{\partial^2 \Pi_k\left[s'_k(s_{-k}), s_{-k}\right]/\partial s_k \partial s_{-k}}{\partial^2 \Pi_k\left[s'_k(s_{-k}), s_{-k}\right]/\partial s_k^2} \tag{3.81}$$

which is negative (positive) for strategic substitutes (complements). Markets for strategic substitutes are markets in which firms set quantities, markets for strategic complements are markets in which firms compete by prices. Hence, the strategy vector is constrained by the condition that a maximum quantity exists for which demand is saturated (strategic substitutes) or by the condition that a maximum willingness to pay exists by which demand is choked off (strategic complements). Condition (3.82) guarantees that the standard comparative statics results hold:

$$\frac{ds'_k}{ds_{-k}} \frac{ds'_{-k}}{ds_k} < 1. \tag{3.82}$$

The preceding sections have employed a specific model of orthogonal reaction functions. Orthogonal reaction functions result from a payoff function for which $\partial^2 \Pi_k/\partial s_k \partial s_{-k} = 0$. This specification is not applicable to industrial markets. However, this section will show that these results carry over to models which imply non-orthogonal reaction functions.

Although the general shape of the constraints is unknown for non-orthogonal reaction functions, one may determine which constraint is binding when two firms want to move away from the non-cooperative equilibrium. It is obvious that all three compliance constraints are equal to zero at the non-cooperative strategy levels. Additionally, it has been demonstrated that only the *ex ante* and the *ex post* compliance constraints are relevant because both imply (2.16). Hence, different slopes of the *ex ante* and the *ex post* compliance constraint for this point mean that one constraint dominates the other constraint. The slopes of all compliance constraints are given by (3.83), (3.84) and (3.85):

$$\left[\frac{ds_{-k}}{ds_k}(\cdot)\right]_{CC_k^1 = 0} = 0, \tag{3.83}$$

$$\left[\frac{ds_{-k}(\cdot)}{ds_k}\right]_{CC_k^2=0} = \frac{ds'_{-k}}{ds_k}, \qquad (3.84)$$

$$\left[\frac{ds_{-k}(\cdot)}{ds_k}\right]_{CC_k^3=0} = -\frac{1-\delta_k^n}{\delta_k^n}\frac{ds'_{-k}}{ds_k}. \qquad (3.85)$$

Equations (3.83), (3.84) and (3.85) give the slopes of the compliance constraints (2.16), (2.17) and (2.19) at the non-cooperative strategy levels. They show that the slope of the compliance constraints differ at the non-cooperative strategy level. For the case of industrial markets, mutual improvement of both firms implies that ds_{-k}/ds_k should be positive: in a market for strategic substitutes, both firms gain mutually only by jointly reducing the supplied quantities, in a market for strategic complements, both firms gain mutually only by increasing the charged prices. Figure 3.10 shows the respective reaction curves and iso-profit curves.

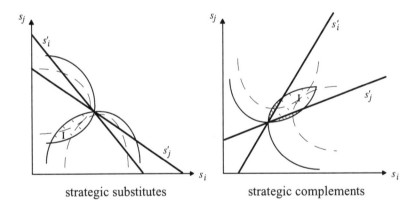

strategic substitutes	strategic complements

Figure 3.10   Non-cooperative equilibrium and mutual improvement in duopolistic markets

The area which lies between both firms' iso-profit curves at the non-cooperative equilibrium (denoted by I) is the area of mutual improvement. It is restricted further by the compliance constraints. The iso-profit curves at the non-cooperative equilibrium are given by the unbroken line, and iso-profit curves which give higher payoffs are given by broken lines. In the case of strategic substitutes, firm i (j) improves its profits by shifting its iso-profit curves downwards (leftwards) because a decrease in s_j (s_i) (that is, the supply of its rival) for a constant s_i (s_j) (that is, a constant own supply) benefits firm i (j). In the case of strategic complements, firm i (j) improves its profits by shifting its iso-profit curves upwards (leftwards) because an increase in s_j (s_i) (that is, the price of its rival) for a constant s_i (s_j) (that is, a constant own price) benefits firm i (j). Since the broken iso-profit curves are tangential at a point in I in both diagrams, the tangential points represent payoff and strategy combinations on the contract curve. Figure 3.10 depicts linear reaction curves for the sake of simplicity. As the slope of the reaction curve is negative for strategic substitutes and positive for strategic complements, (3.85) is the binding constraint for strategic substitutes and (3.84) is the binding constraint for strategic complements. Figure 3.10 shows that strategic complements allow mutual improvements to be realized which give only one firm an incentive to defect because the area of mutual improvements includes a part of each firm's reaction curve. An outcome on a firm's reaction curve inside I benefits both firms but makes defection profitable in one period only for one firm. The shaded area gives outcomes for which defection of one firm raises no problem. Equation (3.84) gives the binding constraint around the non-cooperative equilibrium because the slope of (3.85) is negative. Since s'_i is steeper than s'_j in order to guarantee the standard comparative statics results (see (3.82)), there is always scope for mutual improvement.

This result does not hold for strategic substitutes because improvement on the non-cooperative outcome makes defection profitable for both firms. Additionally, it is not guaranteed that the compliance constraints allow mutual improvements to be realized in every case. Mutual improvements from this point are ruled out if the slope of CC_i^3 is flatter than the slope of CC_j^3 because it would require strategy levels to meet agent i's compliance constraint which violate agent j's compliance constraint. In this case, firm i (j) has to choose strategy levels in order to guarantee compliance of firm j (i) which makes defection of firm i (j) profitable in the long run. Figure 3.11 shows the two possibilities. CC_i^3 gives the maximum s_j as a function of s_i for which the compliance of firm i is guaranteed, CC_j^3 gives the maximum s_i as a function of s_j for which the compliance of firm j is guaranteed.

Figure 3.11 depicts the slope of the relevant compliance constraint at the non-cooperative strategy level. The general shape of the compliance constraints is unknown. Since more cooperation makes defection more profitable, one may suppose that the constraints are more demanding when more cooperation is to be sustained. In this case, the strategy level of firm j (i) has to decrease overproportionally with a decreased strategy level of firm i (j) in order to guarantee compliance of firm i (j). If the compliance constraints are more demanding when both firms' profits are increased beyond the non-cooperative levels, no scope for mutual improvement at the non-cooperative Nash equilibrium implies no scope for mutual improvement at all.

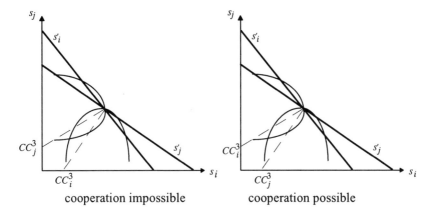

Figure 3.11 Scope for cooperation for strategic substitutes

In this case, strategic substitutes require

$$\frac{ds'_j}{ds_i}(\cdot)\frac{ds'_i}{ds_j}(\cdot) < \frac{\delta_i^n \delta_j^n}{\left(1-\delta_i^n\right)\left(1-\delta_j^n\right)} \tag{3.86}$$

in order to be able to realize mutual improvement. Condition (3.86) specifies that the slope of CC_i^3 falls short of the slope of CC_j^3. Condition (3.86) is always fulfilled for orthogonal reaction functions.

It should be recalled and stressed that any general model must assume a concave and at least piecewise differentiable payoff frontier which gives the set of efficient agreements. The preceding sections have demonstrated that solutions on a compliance constraint are relevant. This result holds for industrial markets as well. Suppose that any firm is able to exert such a strong bargaining power that it is able to realize its optimal policy. If the compliance constraints were irrelevant, its best policy would maximize its profits subject to giving the other firm its non-cooperative profits. This policy, however, would ensure the defection of the other firm. Hence, the optimal policy has to be found on a compliance constraint because overfulfilment of all compliance constraints would leave bargaining gains unexploited. If the payoff frontier is concave and at least piecewise differentiable, ρ is well defined in certain ranges and discount factor combinations exist which specify an agreement on a compliance constraint.

It can also be shown that other basic results of this chapter carry over to duopolistic markets. The compliance effect can be made clear by recalling (2.27), (2.28) and (2.29):

$$U_k^1(\tilde{n}, \delta_k) := (1 - \delta_k) U_k',$$
(2.27)

$$U_k^2(\tilde{n}, \delta_k) := \frac{1 - \delta_k}{1 - \delta_k^{\tilde{n}+1}} U_k' + \frac{\delta_k \left[1 - \delta_k^{\tilde{n}}\right]}{1 - \delta_k^{\tilde{n}+1}} U_k^p,$$
(2.28)

$$U_k^3(\tilde{n}, \delta_k) := -\frac{1 - \delta_k^{\tilde{n}}}{\delta_k^{\tilde{n}}} U_k^p.$$
(2.29)

Equations (2.27), (2.28) and (2.29) give the critical bargaining gains for firm k which are necessary to make firm k indifferent between defection and compliance. For a discount factor which falls short of one, all critical levels must be strictly positive. This means that a duopolist cannot realize his/her optimal policies without guaranteeing positive bargaining gains for his/her opponent. If contracts were enforceable, every optimal policy would leave the other firm on its iso-profit curve of the non-cooperative equilibrium. As this policy would lead to non-compliance, it is clear that the optimal policy sets (2.28) or (2.29) equal to zero.

Additionally, (2.30), (2.31) and (2.32) have demonstrated that the critical levels decrease with the discount factor. This effect is precisely the compliance effect: if the discount factor of a certain firm decreases and the solution has to be found on either the *ex ante* or the *ex post* compliance constraint defined by this firm, compliance can only be guaranteed by higher bargaining gains. The bargaining effect goes in the opposite direction. Since the total effect was already ambiguous for orthogonal reaction curves, it is obviously also ambiguous for non-orthogonal reaction curves.

This result gives a remarkable interpretation for implicit collusion between two suppliers when the capital structure is changed for one firm. Suppose that in the beginning both firms were owned by families who also manage the firms. This capital structure may lead to a high degree of institutional toughness because the future has a strong weight for managers who are likely to bequeath the firm to their children. Now suppose that one firm will be going public, and it will no longer be managed by family members. Going public gives the shareholder value a stronger weight and, as the success of the manager depends on the firm's share price which itself depends on the current performance of the firm, the discount factor describing the weight of future payoffs for current decision may be decreased. If this firm had a self-enforcing contract with the other firm, it is not known whether this changed capital structure will decrease or increase its payoffs.

In general, it is unknown which compliance constraint is binding but certain specifications allow the dominating constraint to be worked out.

Consider, for example, a duopolistic market for strategic substitutes. Market demand can be given by a linear inverse demand function:

$$p = a - b\left[s_i + s_j\right].$$

(3.87)

p denotes the equilibrium price, a denotes the reservation price, b is the slope of the inverse demand function, and s_i and s_j are the strategy levels which were already introduced. These strategy levels are quantities, and since they have the same influence on the price, this duopoly is a typical Cournot duopoly. Suppose further that costs are linear with respect to individual output:

$$\forall k \in \{i, j\}: C_k = c_k s_k \text{ with } a > \max\{c_i, c_j\}.$$

(3.88)

The condition for a ensures that the market will be served by at least one firm. The individual profits are

$$\Pi_k\left[s_k, s_{-k}\right] = \left\{a - b\left[s_k + s_{-k}\right]\right\}s_k - c_k s_k.$$

(3.89)

Maximization of (3.89) with respect to s_k gives the reaction function:

$$s_k'(s_{-k}) = \frac{a - b s_{-k} - c_k}{2b} \text{ with } \frac{ds_k'}{ds_{-k}}(s_{-k}) = -\frac{1}{2}.$$

(3.90)

The reaction function (3.90) determines the non-cooperative strategy levels and the non-cooperative profits:

$$s_k^n = \frac{a + c_{-k} - 2c_k}{3b}, \quad \Pi_k^n = \frac{(a + c_{-k} - 2c_k)^2}{9b}.$$

(3.91)

The non-cooperative levels are denoted by the superscript n. Equations (3.90) and (3.91) assume that it does not pay for the low-cost firm to serve the whole market alone. Let c_{max} (c_{min}) denote the higher (lower) unit costs and let l represent the low-cost firm. If (3.92) is fulfilled, it does not pay for the low-cost firm to serve the whole market by charging the unit costs of the other firm:

$$\left[\frac{a-c_{max}}{b}-c_{min}\right]\frac{a-c_{max}}{b}\leq\frac{\left(a+c_{-l}-2c_l\right)^2}{9b}. \tag{3.92}$$

Condition (3.92) rules out that limit pricing is profitable. The bargaining gains, the profits of unilateral defection, and the profits of investing in a resumption cooperation for n periods are relevant for all compliance constraints. The profits of defection can be derived by the use of the reaction function:

$$\Pi'_k\left(s_{-k}\right)=\frac{\left(a-bs_{-k}-c_k\right)^2}{4b} \quad \text{with} \quad \frac{d\Pi'_k}{ds_{-k}}\left(s_{-k}\right)=-\frac{a-bs_j-c_i}{2}. \tag{3.93}$$

The defection profits will be denoted by a prime and are a function of the other firm's strategy level. The profits of investing in a resumption of cooperation depend on the own strategy level because the other firm will react along its reaction curve during the punishment phase. These profits are denoted by the superscript p, and they can be derived by inserting the other firm's reaction function into the profit function of the firm under consideration:

$$\Pi^p_k\left(s_k\right)=\frac{a-bs_k-2c_k+c_{-k}}{2}s_k \quad \text{with} \quad \frac{d\Pi^p_k}{ds_k}\left(s_k\right)=\frac{a-2bs_k-2c_k+c_{-k}}{2} \tag{3.94}$$

This section has shown that the *ex post* compliance constraint is binding at the non-cooperative equilibrium when firms compete using strategic substitutes. This result can be used to demonstrate that the *ex post* compliance constraint always dominates the *ex ante* compliance constraint.

The way one arrives at this result is to consider the slope of the first two compliance constraints in general. Differentiating (3.89) and using (3.93) and (3.94) gives

$$\left[\frac{ds_j}{ds_i}\right]_{CC^1_i=0}=\frac{a-2bs_i-bs_j-c_i}{bs_i-\left(1-\delta_i\right)\left(a-bs_j-c_i\right)/2} \tag{3.95}$$

and

$$\left[\frac{ds_j}{ds_i}\right]_{CC_i^2=0} = \frac{a - 2bs_i - bs_j - c_i + \delta_i\left(1 - \delta_i^n\right)s_i'\Big/3\left(1 - \delta_i^{n+1}\right)}{bs_i - \left(1 - \delta_i\right)\left(a - bs_j - c_i\right)\Big/2\left(1 - \delta_i^{n+1}\right)}. \tag{3.96}$$

The first derivative gives the slope for the condition which is due to the trigger strategy, the second derivative gives the slope for the *ex ante* compliance constraint. Comparing both terms reveals that the numerator of (3.95) is always smaller than the numerator of (3.96) (see the term containing s_i'), and that the denominator of (3.95) is always greater than the denominator of (3.96). Since it has been shown that CC_i^2 and CC_i^3 imply CC_i^1, it is clear that all compliance constraints intersect at the same point in the s_i–s_j space when they intersect. Then, (3.95) and (3.96) demonstrate that these two compliance constraints cannot intersect except in the non-cooperative equilibrium. If they can never intersect, the other compliance constraint can also never intersect. Hence, they intersect only once, and since the *ex post* compliance constraint dominates at the non-cooperative equilibrium for strategic substitutes, it dominates all other compliance constraints for all relevant strategy levels. This constraint would lose its dominance if and only if all compliance constraints intersect a second time.

One may conclude that the *ex post* compliance constraint is the only relevant constraint for a Cournot duopoly model under constant unit costs and linear demand. In Section 2.4 , it was demonstrated that the scope for cooperation depends on n as well. A low n makes the *ex post* compliance constraint less demanding. As it has been shown that the *ex post* compliance constraint always dominates, the scope for cooperation is maximized by setting n equal to one. Hence, strategic bargaining for implicit collusion between two quantity-setting duopolists with constant unit costs can be expected to specify the shortest punishment phase. Inserting (3.89), (3.91), (3.93) and (3.94) for $n = 1$ into (2.21) gives the compliance of firm i:

$$G[s_j, c_i] := \left[a - c_i - b\left(s_i + s_j\right)\right]s_i - \frac{1}{\delta_i}\frac{\left(a + c_j - 2c_i\right)^2}{9b}$$
$$+ \frac{1 - \delta_i}{2\delta_i}\left[a - bs_i - 2c_i + c_j\right]s_i. \tag{3.97}$$

Definition (3.97) gives *ex post* compliance not in terms of bargaining gains but in terms of total profits. If an agreement has to be found on firm i's compliance constraint, the possible strategy combinations are given by

(3.98). The functional specification allows the effect of a change in firm i's production costs on the necessary strategy level of firm j to be considered:

$$\frac{\partial G}{\partial s_j}(\cdot) = -bs_i < 0,$$

$$\frac{\partial G}{\partial c_i}(\cdot) = \frac{1}{\delta_i}\left[\frac{2(a+c_j-2c_i)}{9b} - s_i\right] \begin{array}{c}< \\ > \end{array} 0 \Leftrightarrow s_i \begin{array}{c}> \\ < \end{array} \frac{4\left[a+c_j-2c_i\right]}{9b}$$

$$\Rightarrow \frac{ds_j}{dc_i} > 0 \text{ because } s_i \leq s_i^n = \frac{a+c_j-2c_i}{3b}. \tag{3.98}$$

Equation (3.98) demonstrates that – all other terms unchanged – a firm may increase its strategy level in order to guarantee compliance of the other firm when the firm which defines the compliance constraint reduces its unit costs. An increase in the strategy level in a market for strategic substitutes would mean an increase in profits. The critical s_i is so high that $ds_j/dc_i > 0$ is guaranteed in all cases because $ds_j/dc_i < 0$ would require an individual supply of firm i above its non-cooperative level. This requirement contradicts the requirement of mutual improvement of both firms. Hence, firm j has to sacrifice fewer compliance costs when the unit costs of firm i decrease. In this case, it may react to reduced unit costs by an increase in output and profits. This is of course only one effect because a change in c_i changes the strategic bargaining solution and hence s_i as well. But it demonstrates that lower costs of a firm may reduce the compliance costs of the other firm *ceteris paribus*.

3.6 SUMMARY

If transfers cannot support cooperation in a non-cooperative environment, strategic bargaining for a self-enforcing contract defines a bargaining problem which cannot decouple individual effort determination and the distribution of bargaining gains. This fact implies several results which shape strategic bargaining for a self-enforcing contract without transfers:

- The optimal policies of each agent do not specify an agreement on the contract curve but on at least one compliance constraint curve.

    If one agent realizes maximum bargaining gains because he/she is perfectly patient, he/she deliberately agrees to an outcome which is not Pareto efficient in the usual sense. The reason is that the other agent must be given some bargaining gains compared to no cooperation in order to ensure his/her compliance. This result implies that a duopolist's best policy does not leave the other duopolist on his/her non-cooperative profit level but increases his/her profits in order to make defection from implicit collusion unprofitable.

- If both agents are extremely impatient because δ_i, $\delta_j < 0.25$ holds and utilities are given by (3.1), both agents' optimal policies coincide and are determined by the intersection of the compliance constraint curves. In this case, there is no bargaining because all other feasible agreements are Pareto dominated by this agreement.

Strategic bargaining is trivial when there is no conflict between the bargaining agents because one feasible agreement Pareto dominates all other feasible agreements. This case is given if both agents' discount factors are very low. Very low discount factors may result from very long periods between realizations. If chances for implicit collusion are repeated, say, every ten years, one may expect that discount factors are to be found in this range. Conversely, one may expect that the set of potential self-enforcing contracts is no singleton if repetition occurs without significantly long delays.

- If neither agent is extremely impatient, different Pareto-unranked agreements exist which can be described by the concession line. The concession line which gives all candidates for the strategic bargaining solution includes parts of the compliance constraint curves but may exclude the contract curve completely. If the contract curve is not completely dominated by the compliance constraints, different discount factor combinations imply a unique bargaining solution and vice versa. If the contract curve is completely dominated, different discount factor combinations may imply the same bargaining solution.

For the case of orthogonal reaction functions, the chapter has demonstrated that the development of the concession line is straightforward because all compliance constraints of each agent coincide for a punishment period of unity.

- Strategic bargaining may specify a self-enforcing contract which lies on a compliance constraint curve even if agreements are feasible on the contract curve.

This implication may be found to represent the salient result of strategic bargaining for self-enforcing contracts without transfers: even if agreements which are Pareto optimal in the usual sense can be supported in a non-cooperative environment, they may not be the bargaining solution. Instead, the solution may be found on an agent's compliance constraint curve. Thus, it is a serious misunderstanding to focus exclusively on the contract curve when discussing cooperation in a non-cooperative environment. Solutions on an agent's compliance constraint curve may represent the result of rational strategic bargaining subject to potential delay costs between two agents. Hence, any cooperation which is observable does not automatically qualify for Pareto optimality in the usual sense even if Pareto-optimal outcomes in this sense could be sustained by repetition.

- Increasing (decreasing) an agent's discount factor increases (decreases) his/her bargaining gains if this discount factor does not define the part of the concession line which determines the strategic bargaining solution. If the discount factor which determines the relevant part of the concession line is changed, the result is ambiguous because two effects have opposite implications: the bargaining effect increases the bargaining power of the relatively patient agent; the compliance effect increases (decreases) the efforts of the other agent which are necessary to guarantee compliance if the discount factor is decreased (increased).

The ambiguous impact of an agent's discount factor sets the stage for discussing team selection. If cooperation is only possible between two partners, each partner must balance the bargaining effect and the compliance effect implied by different partners. For example, a duopolist will not want to join a partnership for implicit collusion with a very impatient partner because his/her costs for guaranteeing the compliance of this partner are very high. Instead, he/she will seek cooperation with a more patient partner whose discount factor balances bargaining and compliance effects.

4. Bargaining for a Collective Good with Transfers

4.1 INTRODUCTION: SELF-ENFORCING COOPERATION WITH TRANSFERS

The previous chapter applied the strategic bargaining model to problems which do not allow for any side payments. This chapter takes the opposite position and assumes that both agents may use transfers to support cooperative outcomes. No institution or agency is either able or intends to restrict transfer policies. Although transfers may serve for shifting realized bargaining gains to the other agent, this chapter will demonstrate that transfers do not necessarily improve on the corresponding outcome without transfers.

One of the applications of the previous chapter was duopolistic markets. If authority control over firms' resource use were not possible, transfer policy could play a role in duopolistic markets as well. If transfers which are not market based cannot be verified as restricting competition by an antitrust board, authority control is restricted to avoiding merger. Merger, however, can only be controlled if the financial and institutional interdependencies among firms can be uncovered. Governmental control which can uncover capital structures and cross-ownerships but cannot uncover transfers which redistribute only bargaining gains from implicit collusion would obviously define a rather strange assumption. Therefore, the model to be developed in this chapter should be applied to issues on which cooperation is unanimously appreciated. If cooperation is appreciated, enforceability is in doubt if the institutional setting is not able to ensure credible commitments of both agents. Compared to the case without transfers, the policy design tries to support but cannot guarantee cooperation by transfers whereas regulating authorities try to restrict cooperation in duopolistic markets in order to avoid welfare losses on the consumer side.

The prototype of an incomplete institutional setting not able to guarantee enforceability is all variants of international relations. No country is able to commit itself credibly to a certain policy because sovereignty enables every country to withdraw from every international agreement whenever it wants to. If a country agrees to pursue a specific policy, this intention is credible only if either the benefits of a contract breach are less than the compliance benefits or sovereignty is restricted by international institutions which can enforce international contracts and which dominate national law.

This chapter addresses the case that there is no restriction to an agent's action: both agents are completely sovereign with respect to their decisions including transfer policies, and no third party is involved in enforcing an agreement. The prominent applications in international relations are international environmental policies and trade policies. There are also other applications, such as international project management, but these applications do not deal with a collective good problem. They reflect a buyer–seller relationship and will be discussed in Chapter 5.

International environmental problems are often due to transboundary spillovers of national environmental resource uses. If two countries' national policies contribute to an environmental public bad from which both countries suffer, both countries are better off by restricting environmental resource use compared to the non-cooperative outcome. A coordination problem arises because every country is even better off if only the other country restricts its resource use. In this case, restricting national resource uses is a collective good as both countries benefit in terms of environmental quality but only one country has to carry the corresponding costs in terms of less production or increased abatement costs.

Meanwhile, the literature on international environmental problems is very extensive (for an overview, see Siebert 1998). Aspects of international coordination, however, have been dealt with mainly by comparing non-cooperative and cooperative behaviour. Cooperative behaviour maximizes both countries' net benefits from joint environmental policies. Maximizing joint benefits, however, assumes implicitly that allocating environmental protection and distributing bargaining gains can be separated. If bilateral environmental treaties must be self-enforcing, this is a strong assumption which will be shown not to be fulfilled in general.

The structure of the model of this chapter is also based on repetition. Obviously, international environmental problems do not always fit this assumption. Pollution problems, for example, arise in many cases because the stock of accumulated pollution and not current pollution endangers environmental quality. Dynamic games of intertemporal resource uses, however, already increase the complexity for the non-cooperative outputs because their solution is determined by whether both countries' strategies depend only on time (open-loop concept) or depend also on some stock variables (closed-loop concept, for an application to greenhouse policies, see Cesar, 1994). Additionally, Section 2.2 has demonstrated that the Folk

Theorem does not hold for dynamic games because the chances of cooperation do not necessarily increase with both agents' discount factors. Chapter 6 will address this point when discussing an agenda for future research.

Adopting the simplest variant of dynamic games, that is, the repeated-game approach, may contribute to a first understanding of international environmental policies in a non-cooperative environment without overwhelming the environmental part with several intertemporal links. Thereby, the model of this chapter can explain why such treaties are often incomplete, that is, they do not prescribe an intended behaviour for all events, when compared with cooperative behaviour. No country is interested in maximizing joint benefits but in maximizing its own welfare subject to guaranteed compliance of the other country. The model will assume that transfer policies must be self-enforcing as well. On this basis, incomplete cooperation will not be surprising. Incomplete cooperation can be explained by binding compliance constraints which also take defection from agreed-upon transfer policies into account.

The same assumption applies to trade policies. Although trade theory has stressed the gains from free trade, trade policy instruments are still in use. In the history of trade policies, non-tariff barriers, such as product standards, import quotas and voluntary export restraints, have replaced tariffs which were the dominant trade policy instruments at the beginning of the internationalization of markets. Enforceability of free trade faces two important restrictions which require self-enforceability of trade agreements: on the one hand, sovereignty constraints restrict the policy options of international institutions (such as the World Trade Organization) to enforce trade agreements. On the other hand, trade agreements are always incomplete because national policy intentions are not so clear-cut that the set of trade policy instruments is well defined. For example, introducing a certain product quality norm could be claimed to protect consumers although actually it is intended to restrict imports (for a striking empirical example, see the German *Reinheitsgebot* as a quality norm regulating the ingredients permitted in the brewing of beer for the German market, which was found to conflict with the rules of the European Common Market). Therefore, it is often not clear whether a country has violated a certain term of a trade agreement or aims at other national policy objectives.

The incentive structure for trade partners is obvious and holds for different theoretical approaches in their standard cases: compared to autarky, both partners gain from freeing trade but one partner may gain even more if he/she introduces trade policy instruments but the other partner does not. In a Heckscher–Ohlin setting, one partner maximizes his/her bargaining gains by imposing an optimal tariff in order to increase his/her terms of trade. In the standard cases, the other partner loses compared to free trade but gains compared to no trade according to the Free-trade-for-one Theorem. If both partners introduce tariffs, they are worse off compared to a free trade situation. In a duopolistic setting with one firm in each country and both producing for a third country's market, strategic trade theory has demonstrated that unilaterally subsidizing production or R&D may shift monopoly rents into the subsidizing country (Brander and Spencer, 1985). If both countries subsidize in order to maximize the profits of their domestic producers, they are worse off compared to a situation in which they bilaterally refrain from giving subsidies. In both cases, a Pareto-dominated outcome gives the non-cooperative solution.

The position taken here assumes that trade policies cannot rely on external enforcement because countries are sovereign and trade policy instruments which are not covered explicitly by an agreement are easy to invent. Two countries agreeing on a certain degree of trade liberalization know that either of them is able to break the self-enforcing part of the contract without violating the terms of the explicit contract. As there is no reason to suppose a certain mover structure, both countries are supposed to decide simultaneously whether they comply with the contract or violate the agreement unilaterally (even if they do not take an action which is obviously ruled out by the literal text of the agreement). A certain trade agreement is therefore self-enforcing if the conditions of weak renegotiation-proofness make a country refrain from the unilateral introduction of additional trade restrictions and make it abolish these trade barriers unilaterally in order to resume cooperation after it has defected.

The most convenient approach for dealing with different trade policy options is to assume variable degrees of liberalization. As in Chapter 3, the degrees will be given in terms of efforts and a high effort level corresponds to a high degree of domestic liberalization. Domestic liberalization will be assumed to benefit both countries but will also be assumed to imply domestic costs. These costs can be given a twofold interpretation. First, liberalization implies opportunity costs, because increasing the terms of trade or shifting monopoly rents is restricted. Second, free trade is likely to change income distribution patterns. In a Heckscher–Ohlin framework, the real income of the relatively scarce factor is decreased by an increased liberalization. This is the well-known Stolper–Samuelson Theorem (see any trade theory textbook, for example, Gandolfo, 1986). In the case of sector-specific factors, free trade decreases the rents of factors in shrinking sectors. If redistributing gains from trade is impossible, the potential loss of firms and sectors may be expected to put political pressure on a government pursuing liberalization policies. If these groups are relatively better organized than the winners from free trade, an increased level of liberalization may imply increasing political costs for the government (for the political economy of protection, see Hillman, 1989). Interest groups support politicians who aim to give these industries a protective shelter, or they disseminate selective information on trade policy issues in order to reduce the public's acceptance of free trade. Lobbying models indicate that free trade does not give an optimum for the so-called political support functions which define the government's objective when lobbying is relevant (see Grossman and Helpman, 1994). Because gains and losses from trade policies are weighted differently due to influential groups, political support for trade policies is maximized rather than the gains from trade.

A model by Grossman and Helpman (1995) which incorporates lobbying compares the non-cooperative solution with a bargaining-based cooperative one but does not address enforceability aspects. The following sections will deal with these aspects explicitly and readers interested in trade policy should interpret the non-cooperative one-shot equilibrium as the solution which maximizes every country's welfare independently by using the whole array of trade policy instruments. The model fits a Heckscher–Ohlin framework, a strategic trade policy and a lobbying setting (including all possible combinations). Of course, transfers which shift gains from trade from one country to another must be self-enforcing as well.

As it is sometimes hard to relate all the ideas of the following sections to concrete applications, this chapter will return to international environmental problems and trade policies in separate sections. International environmental problems will be discussed on the basis of some simulations which use estimated cost and benefit functions for CO_2 reduction policies. Trade policy issues will be discussed by considering a model which involves two countries which have to coordinate their domestic tariff policies using a self-enforcing contract. Accordingly, this chapter is organized as follows. Section 4.2 deals with the role of transfers for the compliance constraints. It demonstrates that transfers double the number of compliance constraints because the non-compliance options differ between a donor and a receiver. Based on this differentiation, Section 4.3 develops the optimal policies of both agents. It shows that the sign of optimal policy transfers depends on the individual bargaining gains when joint bargaining gains are maximized. Thus, it demonstrates that bargaining for a collective good with transfers may result in twelve combinations of optimal policies for both agents which serve as the starting-points of the bargaining process.

Section 4.4 develops the concession policies and the concession line. Because bargaining for a collective good with transfers involves three variables, concession policies must be explicitly addressed. Section 4.5 deals with feasibility constraints, the strategic bargaining solution and the impact of discount factor changes. It also compares bargaining with and without transfers for the optimal policies of an agent. Section 4.6 contains the simulations for self-enforcing environmental contracts. Section 4.7 discusses the impact of strategic bargaining for a self-enforcing trade policy contract on the chances of an agreement which abolishes all trade barriers. Section 4.8 concludes and summarizes this chapter. Since this chapter will focus on the role which transfers may play, it will employ specific functions as well.

4.2 THE ROLE OF TRANSFERS FOR THE COMPLIANCE CONSTRAINTS: THE RECEIVER'S AND THE DONOR'S CASE

Chapter 3 assumed that cooperation must be based on efforts of the agents only and did not allow for side payments. This section will introduce side payments as a possible supplement to a joint effort policy. It will start with a discussion on the implications of side payments by considering their impact on the agents' compliance constraints. This impact deserves its own section because transfer policies are often dealt with in a completely different way from pure effort policies. This section will demonstrate that transfers play a different role in self-enforcing contracts if they are subject to compliance risks as well.

Starting with a naive assessment of transfers for self-enforcing contracts, one could think of transfers as an instrument to cure compliance problems. For example, if the individual compliance constraints of both agents do not dominate the contract curve, agreements on the contract curve are known to maximize total utility. If transfers enter the agent's utilities by simple addition to effort-based utilities, setting $E_i = (\alpha_i + \alpha_j)/\gamma_i$ and $E_j = (\alpha_i + \alpha_j)/\gamma_j$ gives the maximum total utility. If the compliance constraints allow this effort policy to be realized, why should both agents not realize the maximum bargaining gains and distribute the total utility according to the strategic bargaining solution using transfers? Additionally, if a welfare-improving point on the contract curve is dominated by an agent's compliance constraint, the other agent could pay him/her with by transfers in order to enable another effort policy if the total gains of switching to the welfare-improving point exceed the transfers necessary to pursue another effort policy. Why should both agents not switch to the welfare-improving point?

The underlying assumptions which favour a curing role of transfers for self-enforcing contracts, however, are misspecified in these lines of reasoning, which assume that transfer policies are enforceable whereas joint effort policies are not. If transfer policies are enforceable, a certain distribution of bargaining gains can be credibly agreed upon and transfers can decouple the allocation of efforts and the distribution of bargaining gains. Thus, the conclusion that transfers lead to maximizing total utility relies on the assumption of enforceable transfer policies. The second question sheds some light on the role of transfers for weakening compliance constraints. The argument which was given by answering this question, however, is wrong. Consider, for example, an agent whose compliance constraint defines the relevant range of the concession line without transfers. If the other agent gives him/her a transfer in addition to the unchanged joint effort policy, this agent will break the self-enforcing contract because the transfer increases his/her incentive for non-compliance. This result is due to the simultaneous moves of both agents: both agents decide simultaneously whether they will fulfil or break the contract. An agent who restricts the outcome by his/her compliance constraint will therefore break the agreement because transfers increase his/her short-run gains from defection.

The last argument sets the stage for discussing the role of transfers for the compliance constraints. This chapter does not make the puzzling assumption that joint effort policies must be self-enforcing whereas transfer policies are enforceable. Instead, it assumes that both policies must be self-enforcing and cannot be separated. Additionally, the simultaneous-move assumption still holds and both agents decide simultaneously whether they will fulfil or break the contract. Let transfers given from agent j to agent i be denoted by T. A negative T indicates transfers given from agent i to agent j. Throughout this chapter, transfers are assumed to enter both agents' utility functions by addition to the effort-based utilities because V_i and V_j measure payoffs. With transfers, the utility functions of both agents are given by:

$$V_i = \alpha_i \left(E_i + E_j \right) - \frac{\gamma_i}{2} E_i^2 + T ,$$

$$V_j = \alpha_j \left(E_i + E_j \right) - \frac{\gamma_j}{2} E_j^2 - T . \qquad (4.1)$$

If $T > 0$ ($T < 0$) holds, agent j (agent i) takes the donor's role and agent i (agent j) takes the receiver's role. Computing the compliance constraints, however, is not as straightforward as in Chapter 3 but depends on the sign of T. This dependence can be made clear first by assuming that agent j pays a transfer which agent i receives. What does non-compliance mean in this case? Non-compliance of agent j means that agent j does not hold the promised effort level and does not pay the transfer. Non-compliance of agent i means that agent i does not hold the promised effort level but takes the transfer from agent j. If agent i is assumed to pay agent j, the non-compliance results are turned around: agent i does not hold the promised effort level and does not pay the transfer, and agent j does not hold the promised effort level but takes the transfers. Comparing both cases shows that non-compliance depends on the role of the respective agent because a non-compliant receiver will always take the transfer paid by a compliant donor and a non-compliant donor will always refuse transfer payment. As both agents can be imagined as either receiver or donor, two cases have to be distinguished for every agent: these will be referred to as the receiver's and the donor's case.

As in Section 3.3, weak renegotiation-proofness defines three conditions for every case. The first case is defined by $T \geq 0$ which gives agent j the donor's role and agent i the receiver's role. The second case changes the roles and assumes $T \leq 0$. Every self-enforcing contract must guarantee that the sum of discounted current and future benefits from fulfilling the contract must not fall short of the current benefits from breaking the contract in one period and no cooperation in all following periods:

$T \geq 0$:

$$\frac{1}{1-\delta_i}\left[\alpha_i\left[E_i + E_j\right] - \frac{\gamma_i}{2}E_i^2 + T\right] - \left[\frac{\alpha_i^2}{2\gamma_i} + \alpha_i E_j + T\right]$$

$$- \frac{\delta_i}{1-\delta_i}\left[\frac{\alpha_i^2}{2\gamma_i} + \frac{\alpha_i \alpha_j}{\gamma_j}\right] \geq 0,$$

$$\frac{1}{1-\delta_j}\left[\alpha_j\left[E_i + E_j\right] - \frac{\gamma_j}{2}E_j^2 - T\right] - \left[\frac{\alpha_j^2}{2\gamma_j} + \alpha_j E_i\right]$$

$$- \frac{\delta_j}{1-\delta_j}\left[\frac{\alpha_j^2}{2\gamma_j} + \frac{\alpha_i \alpha_j}{\gamma_i}\right] \geq 0, \qquad\qquad (4.2)$$

$T \leq 0$:

$$\frac{1}{1-\delta_i}\left[\alpha_i\left[E_i + E_j\right] - \frac{\gamma_i}{2}E_i^2 + T\right] - \left[\frac{\alpha_i^2}{2\gamma_i} + \alpha_i E_j\right]$$

$$- \frac{\delta_i}{1-\delta_i}\left[\frac{\alpha_i^2}{2\gamma_i} + \frac{\alpha_i \alpha_j}{\gamma_j}\right] \geq 0,$$

$$\frac{1}{1-\delta_j}\left[\alpha_j\left[E_i + E_j\right] - \frac{\gamma_j}{2}E_j^2 - T\right] - \left[\frac{\alpha_j^2}{2\gamma_j} + \alpha_j E_i - T\right]$$

$$-\frac{\delta_j}{1-\delta_j}\left[\frac{\alpha_j^2}{2\gamma_j}+\frac{\alpha_i\alpha_j}{\gamma_i}\right]\geq 0. \tag{4.3}$$

Condition (4.2) gives the first condition which weak renegotiation-proofness implies for the first case, (4.3) for the second case. The second terms make the difference between donors and receivers. For $T = 0$, (4.2) and (4.3) are identical and coincide with (3.6). Weak renegotiation-proofness was shown to add another two conditions which depend on the length of the punishment period. Section 2.4 has shown that these two conditions make the first one redundant. Instead of adopting the general approach as in Section 3.2, this section sets the punishment period equal to one. This assumption maximizes the scope for cooperation if it lets all conditions coincide and allows a single compliance constraint to be defined for every case. Because of the orthogonality of the reaction functions which still holds for the combined effort-transfer policy, this assumption will be shown to produce this result.

Conditions (4.4), (4.5), (4.6) and (4.7) give the conditions of *ex ante* and *ex post* compliance for both cases. Note that a non-compliant receiver has to invest in a resumption of cooperation by providing his/her agreed-upon effort level unilaterally without receiving transfers, whereas a non-compliant donor has to invest by providing his/her agreed-upon reduction level and by paying transfers to the receiver:

$T \geq 0$:

$$(1+\delta_i)\left\{\alpha_i\left[E_i+E_j\right]-\frac{\gamma_i}{2}E_i^2+T\right\}-\left[\frac{\alpha_i^2}{2\gamma_i}+\alpha_iE_j+T\right]$$

$$-\delta_i\left\{\alpha_i\left[E_i+\frac{\alpha_j}{\gamma_j}\right]-\frac{\gamma_i}{2}E_i^2\right\}\geq 0,$$

$$(1+\delta_j)\left\{\alpha_i\left[E_i+E_j\right]-\frac{\gamma_j}{2}E_j^2-T\right\}-\left[\frac{\alpha_j^2}{2\gamma_j}+\alpha_iE_i\right]$$

$$-\delta_j\left\{\alpha_j\left[\frac{\alpha_i}{\gamma_i}+E_j\right]-\frac{\gamma_j}{2}E_j^2-T\right\}\geq 0, \tag{4.4}$$

$T \leq 0$:

$$(1+\delta_i)\left\{\alpha_i\left[E_i+E_j\right]-\frac{\gamma_i}{2}E_i^2+T\right\}-\left[\frac{\alpha_i^2}{2\gamma_i}+\alpha_i E_j\right]$$

$$-\delta_i\left\{\alpha_i\left[E_i+\frac{\alpha_j}{\gamma_j}\right]-\frac{\gamma_i}{2}E_i^2+T\right\}\geq 0,$$

$$(1+\delta_j)\left\{\alpha_j\left[E_i+E_j\right]-\frac{\gamma_j}{2}E_j^2-T\right\}-\left[\frac{\alpha_j^2}{2\gamma_j}+\alpha_j E_i-T\right]$$

$$-\delta_j\left\{\alpha_j\left[\frac{\alpha_i}{\gamma_i}+E_j\right]-\frac{\gamma_j}{2}E_j^2\right\}\geq 0, \tag{4.5}$$

$T \geq 0$:

$$\alpha_i\left[E_i+\frac{\alpha_j}{\gamma_j}\right]-\frac{\gamma_i}{2}E_i^2+\frac{\delta_i}{1-\delta_i}\left\{\alpha_i\left[E_i+E_j\right]-\frac{\gamma_i}{2}E_i^2+T\right\}$$

$$-\frac{1}{1-\delta_i}\left[\frac{\alpha_i^2}{2\gamma_i}+\frac{\alpha_i\alpha_j}{\gamma_j}\right]\geq 0,$$

$$\alpha_j\left[\frac{\alpha_i}{\gamma_i}+E_j\right]-\frac{\gamma_j}{2}E_j^2-T+\frac{\delta_j}{1-\delta_j}\left\{\alpha_j\left[E_i+E_j\right]-\frac{\gamma_j}{2}E_j^2-T\right\}$$

$$-\frac{1}{1-\delta_j}\left[\frac{\alpha_j^2}{2\gamma_j}+\frac{\alpha_i\alpha_j}{\gamma_i}\right]\geq 0, \tag{4.6}$$

$T \leq 0$:

$$\alpha_i\left[E_i+\frac{\alpha_j}{\gamma_j}\right]-\frac{\gamma_i}{2}E_i^2+T+\frac{\delta_i}{1-\delta_i}\left\{\alpha_i\left[E_i+E_j\right]-\frac{\gamma_i}{2}E_i^2+T\right\}$$

$$-\frac{1}{1-\delta_i}\left[\frac{\alpha_i^2}{2\gamma_i}+\frac{\alpha_i\alpha_j}{\gamma_j}\right]\geq 0,$$

$$\alpha_j\left[\frac{\alpha_i}{\gamma_i}+E_j\right]-\frac{\gamma_j}{2}E_j^2+\frac{\delta_j}{1-\delta_j}\left\{\alpha_j\left[E_i+E_j\right]-\frac{\gamma_j}{2}E_j^2-T\right\}$$

$$-\frac{1}{1-\delta_j}\left[\frac{\alpha_j^2}{2\gamma_j}+\frac{\alpha_i\alpha_j}{\gamma_j}\right]\geq 0. \tag{4.7}$$

Fortunately, (4.2), (4.4), (4.6) and (4.3), (4.5), (4.7), respectively, fall together for the one-period punishment, thereby proving the appropriateness of this assumption. Equation (4.8) summarizes the two sets of compliance constraints under the use of $\varepsilon_i = E_i - \alpha_i/\gamma_i$ and $\varepsilon_j = E_j - \alpha_j/\gamma_j$:

$T \geq 0$:

$$E_j^{C(i+)} = \frac{\alpha_j}{\gamma_j} + \frac{1}{\delta_i}\frac{\gamma_i}{2\alpha_i}\varepsilon_i^2 - \frac{T}{\alpha_i},$$

$$E_i^{C(j-)} = \frac{\alpha_i}{\gamma_i} + \frac{1}{\delta_j}\frac{\gamma_j}{2\alpha_j}\varepsilon_j^2 + \frac{1}{\delta_j}\frac{T}{\alpha_j},$$

$T \leq 0$:

$$E_j^{C(i-)} = \frac{\alpha_j}{\gamma_j} + \frac{1}{\delta_i}\frac{\gamma_i}{2\alpha_i}\varepsilon_i^2 - \frac{1}{\delta_i}\frac{T}{\alpha_i},$$

$$E_i^{C(j+)} = \frac{\alpha_i}{\gamma_i} + \frac{1}{\delta_j}\frac{\gamma_j}{2\alpha_j}\varepsilon_j^2 + \frac{T}{\alpha_j}. \tag{4.8}$$

The superscripts $C(i)$ and $C(j)$ denote the compliance constraints which agent i and agent j impose on the other agent's (that is, j's or i's) efforts. The supplement $+$ $(-)$ indicates that the respective agent is a receiver (donor).

Since the compliance constraint curves are likely to define a certain range of the concession line, (4.8) shows that transfers increase complexity substantially, which is also demonstrated by Figure 4.1.

Figure 4.1 gives the compliance constraints for three cases: with a negative T, a positive T and a zero T. The respective compliance constraints are indicated by the superscripts which (3.12) and (4.8) have introduced. A negative T shifts the zero-T compliance constraint defined by agent i upwards and the compliance constraint defined by agent j leftwards. Figure 4.1 assumes that $\alpha_i = \alpha_j$ holds. In this case, the upward shift along the ordinate must exceed the leftward shift along the abscissa because $(1/\delta_i)(T/\alpha_i)$ exceeds T/α_j in this case (see (4.8)). *Ceteris paribus*, introducing transfers from agent j to agent i induces an increase in the minimum E_j and a decrease in the minimum E_i which both guarantee compliance of the other agent. For $\alpha_i = \alpha_j$, the increase in E_j exceeds the absolute value of the decrease in E_i when agent j pays strictly positive transfers to agent i (see (4.8)).

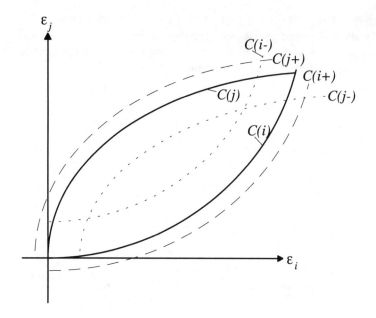

Figure 4.1   Compliance constraints for different transfer levels

A positive T shifts the zero-T compliance constraint defined by agent i downwards and the compliance constraint defined by agent j rightwards. For $\alpha_i = \alpha_j$ the absolute value of the decrease in the minimum E_j (guaranteeing compliance of i) falls short of the increase in the minimum E_i (guaranteeing compliance of j). Figure 4.2 reproduces the case of a zero T and a positive T. Additionally, the utility reservation curves are drawn for both agents.

The utility reservation curves are indicated by $R(i)$ and $R(j)$ for the zero-T case (see also (3.15)) and by $R(i+)$ and $R(j-)$ for the positive case. These functions do not depend on the sign of T and are given in general by

$$E_j^{R(i\cdot)} = \frac{\alpha_j}{\gamma_j} + \frac{\gamma_i}{2\alpha_i}\varepsilon_i^2 - \frac{T}{\alpha_i},$$

$$E_i^{R(j\cdot)} = \frac{\alpha_i}{\gamma_i} + \frac{\gamma_i}{2\alpha_j}\varepsilon_j^2 + \frac{T}{\alpha_j}. \tag{4.9}$$

The superscripts $R(i\cdot)$ and $R(j\cdot)$ denote the necessary effort level of agent j and agent i in order to guarantee the non-cooperative utility level of agent i and agent j, respectively. Figure 4.2 shows that $C(i+)$ and $R(i+)$ have the same starting-point on the ordinate. $C(j-)$ starts on the RHS of $R(j-)$ on the abscissa. This means that a transfer from agent *j* to agent *i* requires that agent *i* provides a minimum effort level which exceeds his/her non-cooperative one even if agent *j* is still at his/her non-cooperative level. Otherwise, agent *j* (the donor) would break the contract. This minimum effort level has to exceed not only the utility reservation level but even a higher level which is defined by agent *j*'s discount factor. Note that transfers enable the donor to substitute for his/her own efforts. An increase in transfers relaxes the pressure on his/her minimum efforts. Thus, transfers may have a beneficial impact on an individual compliance constraint.

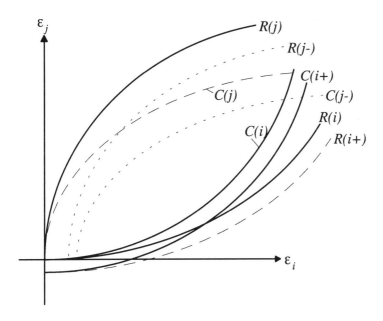

Figure 4.2   *Utility reservation curves and compliance constraints for different transfer levels*

4.3 OPTIMAL POLICIES

When the compliance constraints depend on the sign of a strategic variable, optimal policies depend on the sign of transfers, too. This section will determine optimal policies for an agent and their dependence on specific parameters of the model. First, it will assume non-positive or non-negative transfers. Determining the optimality conditions gives the link to the scope of cooperation allowed by transfers. This section will demonstrate that the sign of optimal transfer policies of an agent, that is, whether his/her optimal policy specifies whether he/she will receive or pay transfers, depends on the sign of the other agent's bargaining gains for $E_i = (\alpha_i + \alpha_j)/\gamma_i$ and $E_j = (\alpha_i + \alpha_j)/\gamma_j$ if no transfers are paid. Bargaining gains without transfers may become negative if marginal benefits are low, thereby changing the problem partially from a purely public good problem into a buyer–seller problem. Determining the optimal policies will also serve for the next section because optimal policies define the relevant range of the concession line. This section starts with considering the optimal policies for agent j. Assume that $T \leq 0$ holds, that is, agent j is a receiver and agent i is a donor, and that the compliance constraint of agent j does not bind. An optimal policy of agent j will set the compliance constraint of agent i equal to zero: this condition ensures that agent i introduces a best policy for agent j. Note that this condition was also the basis for determining optimal policies for self-enforcing contracts without transfers. Assuming $T \leq 0$ allows agent j's optimal policies to be computed by solving the maximization problem

$$\max_{E_i, E_j, T} V_j \qquad \text{s.t.} \, T \leq 0. \qquad (4.10)$$

The optimal values can be determined by solving the third line of (4.8) for T and inserting T into V_j. Let T_j denote the transfers received by j, that is,

$$T_j = \delta_i \alpha_i \frac{\alpha_j}{\gamma_j} + \frac{\gamma_i}{2} \left[E_i - \frac{\alpha_i}{\gamma_i} \right]^2 - \delta_i \alpha_i E_j \qquad (4.11)$$

which gives a simple maximization problem in the form of

$$\max_{E_i, E_j} \left\{ \alpha_j \left[E_i + E_j \right] - \frac{\gamma_j}{2} E_j^2 - \delta_i \alpha_i \frac{\alpha_j}{\gamma_j} - \frac{\gamma_i}{2} \left[E_i - \frac{\alpha_i}{\gamma_i} \right]^2 + \delta_i \alpha_i E_j \right\}.$$

$$(4.12)$$

Note that (4.11) and (4.12) do not exclude positive transfers. Thus, the solution of (4.12) must also satisfy $T_j \leq 0$. This section proceeds first, assuming that $T_j \leq 0$ holds for the optimal solution of (4.12) and second, computing the general solution along the lines of the Kuhn–Tucker Theorem. The optimal solutions for (4.12) are denoted by a star:

$$E_i^* = \frac{\alpha_i + \alpha_j}{\gamma_i}, \quad E_j^* = \frac{\delta_i \alpha_i + \alpha_j}{\gamma_j},$$

$$T_j^* = \frac{\alpha_j^2}{2\gamma_i} - \frac{\delta_i^2 \alpha_i^2}{\gamma_j}. \tag{4.13}$$

The indicated T_j^* is only the optimal transfer level if T_j^* is non-positive. In this case, the bargaining gains for both agents which result from the optimal policy for agent j are given by

$$U_i^* = (1 - \delta_i)\frac{\alpha_i^2}{\gamma_j}, \quad U_j^* = \frac{\alpha_j^2}{2\gamma_i} + \frac{\delta_i^2 \alpha_i^2}{2\gamma_j}. \tag{4.14}$$

If T_j^* in (4.13) is positive, (4.13) does not give the optimal solution. Instead, the optimal solution is given by setting T_j^* zero according to the Kuhn–Tucker Theorem. A non-negative T_j^* in (4.13) is given if δ_i is so low that

$$\delta_i \leq \sqrt{\frac{\alpha_j^2 \gamma_j}{2\alpha_i^2 \gamma_i}} \tag{4.15}$$

holds. Condition (4.15) defines the range of discount factors of agent i which imply a zero transfer level for the optimal policies of agent j, given that transfers are non-positive. If (4.15) holds, optimal policies for $T_j \leq 0$ are given by the optimal policies which Section 3.2 evaluated. These results can be summarized by Table 4.1.

Table 4.1 is an extension of Table 3.1 which summarized the optimal policies for the no-transfer case. Note that the ranges which defined the different cases in Table 3.1 do not coincide with the range of the second column of Table 4.1. If (4.15) is violated, the optimal policies are defined by (4.13) and not by Table 3.1.

Table 4.1 Optimal policies for agent j *in the case* $T_j \leq 0$

Range of agent i's discount factor	Optimal policies are defined by
$\delta_i > \sqrt{\dfrac{\alpha_j^2 \gamma_j}{2\alpha_i^2 \gamma_i}}$	(4.13)
$\delta_i \leq \sqrt{\dfrac{\alpha_j^2 \gamma_j}{2\alpha_i^2 \gamma_i}}$	Table 3.1

The results summarized in Table 4.1 are restricted to $T_j \leq 0$. If $T_j \geq 0$ should hold, agent j should be the donor and agent i should be the receiver. For this case, the compliance constraint defined by the receiver i differs from the one the donor i defined. The first line of (4.8) can be rewritten in order to determine T_j which now indicates the transfers paid by agent j:

$$T_j = \frac{\alpha_i \alpha_j}{\gamma_i} + \frac{\gamma_i}{2\delta_i}\left[E_i - \frac{\alpha_i}{\gamma_i}\right]^2 - \alpha_i E_j.$$ (4.16)

The maximization problem:

$$\max_{E_i, E_j, T} V_j \qquad s.t. T \geq 0$$ (4.17)

will be solved first under the assumption that $T_j \geq 0$ holds. This assumption transforms (4.17) into the simple maximization problem

$$\max_{E_i, E_j}\left\{\alpha_j(E_i + E_j) - \frac{\gamma_j}{2}E_j^2 - \frac{\alpha_i \alpha_j}{\gamma_j} - \frac{\gamma_i}{2\delta_i}\left[E_i - \frac{\alpha_i}{\gamma_i}\right]^2 + \alpha_i E_j\right\}.$$

(4.18)

If $T_j \geq 0$ holds, (4.18) implies the optimal solution

$$E_i^* = \frac{\alpha_i + \delta_i \alpha_j}{\gamma_i}, E_j^* = \frac{\alpha_i + \alpha_j}{\gamma_j},$$

$$T_j^* = \frac{\delta_i \alpha_j^2}{2\gamma_i} - \frac{\alpha_i^2}{\gamma_j}. \tag{4.19}$$

As in (4.13), optimal values are denoted by a star. The indicated T_j^* is the optimal transfer level only if T_j^* is non-negative. In this case, the bargaining gains for both agents which result from the optimal policy for agent j are given by

$$U_i^* = \delta_i (1 - \delta_i) \frac{\alpha_j^2}{2\gamma_i}, \ U_j^* = \frac{\delta_i \alpha_j^2}{2\gamma_i} + \frac{\alpha_i^2}{2\gamma_j}. \tag{4.20}$$

If T_j^* in (4.19) is negative, (4.19) does not give the optimal solution which is given by setting T_j^* zero for this case. A non-positive T_j^* is given if δ_i is so low that

$$\delta_i \leq \frac{2\alpha_i^2 \gamma_i}{\alpha_j^2 \gamma_j} \tag{4.21}$$

holds. Condition (4.21) defines the range of discount factors of agent i which imply a zero transfer level for the optimal policies of agent j, given that transfers are non-negative. If (4.21) holds, optimal policies have been evaluated by Section 3.2. These results can be summarized by Table 4.2. The same reservation which was given for Table 4.1 applies to Table 4.2.

Table 4.2 Optimal policies for agent j *in the case* $T_j \geq 0$

Range of agent i's discount factor	Optimal policies are defined by
$\delta_i > \dfrac{2\alpha_i^2 \gamma_i}{\alpha_j^2 \gamma_j}$	(4.19)
$\delta_i \leq \dfrac{2\alpha_i^2 \gamma_i}{\alpha_j^2 \gamma_j}$	Table 3.1

Comparing Table 4.1 with Table 4.2, it is obvious that the definition of optimal policies depends crucially on $(2\alpha_i^2\gamma_i)/(\alpha_j^2\gamma_j)$. This term enters the first column of Table 4.2 directly and the inverse of this term enters the root in the first column of Table 4.1. This term has a straightforward interpretation. Suppose that contracts were enforceable and consider the optimal policies of both agents. Obviously, enforceability can decouple the allocation of efforts and the distribution of bargaining gains. For enforceable contracts, both agents would seek to maximize the sum of bargaining gains and redistribute the gains in line with the strategic bargaining equilibrium. Maximizing the sum of bargaining gains requires an effort policy

$$E_i = \frac{\alpha_i + \alpha_j}{\gamma_i}, \ E_j = \frac{\alpha_i + \alpha_j}{\gamma_j}.$$

These effort levels will be referred to as the Pareto-optimal effort levels. Before the individual bargaining gains are redistributed, both agents realize bargaining gains according to (see Section 3.2)

$$U_i^* = \frac{\alpha_i^2}{\gamma_j} - \frac{\alpha_j^2}{2\gamma_i}, \ U_j^* = \frac{\alpha_j^2}{\gamma_i} - \frac{\alpha_i^2}{2\gamma_j}. \tag{3.4}$$

Equation (3.4) does not exclude the possibility that the individual bargaining gains before transfers are paid are negative for an agent. This case had to be excluded for Chapter 3 but it is possible for the model here. Negative bargaining gains before transfers for the best enforceable contract are due to significant asymmetries with respect to benefits and costs between the agents. For example, low marginal benefits and costs of agent i can produce a negative U_i if agent j's marginal benefits and costs are high. In this case, agent j has a high preference for a joint effort policy whereas the preference of agent i is low and must be increased by transfers. Significant differences between the agents' parameters are therefore apt to change the problem partially from a public goods' problem to a buyer–seller problem.

The sign of U_i in (3.4) corresponds with the term which decides on the optimal policy of agent j:

$$\frac{\alpha_i^2}{\gamma_j} - \frac{\alpha_j^2}{2\gamma_i} \underset{>}{<} 0 \Leftrightarrow \frac{2\alpha_i^2\gamma_i}{\alpha_j^2\gamma_j} \underset{<}{>} 1. \tag{4.22}$$

If $(2\alpha_i^2\gamma_i)/(\alpha_j^2\gamma_j)$ exceeds unity, agent i realizes bargaining gains in the best enforceable contract even without transfers, if $(2\alpha_i^2\gamma_i)/(\alpha_j^2\gamma_j)$ falls short of one, agent i is at least partially a seller of services in the best enforceable contract and must be compensated by transfers in this case. The expression $(2\alpha_i^2\gamma_i)/(\alpha_j^2\gamma_j)$ is an indicator for the type of the problem: in the first case, agent i is an agent mainly contributing to a public good, in the second case, agent i is an agent mainly selling a private good to agent j. These two cases will be referred to as the public good contributor attribute and the seller attribute, respectively.

The expression $(2\alpha_i^2\gamma_i)/(\alpha_j^2\gamma_j)$ can be used to combine Tables 4.1 and 4.2 in order to determine the optimal policies for agent j in general. If $(2\alpha_i^2\gamma_i)/(\alpha_j^2\gamma_j)$ exceeds one, there is no scope for a positive T_j^*. Table 4.2 indicates that δ_i must exceed $(2\alpha_i^2\gamma_i)/(\alpha_j^2\gamma_j)$ for a positive T_j^* which is impossible for $(2\alpha_i^2\gamma_i)/(\alpha_j^2\gamma_j) > 1$. If $(2\alpha_i^2\gamma_i)/(\alpha_j^2\gamma_j)$ falls short of unity, there is no scope for a negative T_j^* because

$$\frac{2\alpha_i^2\gamma_i}{\alpha_j^2\gamma_j} < 1 \Leftrightarrow \frac{\alpha_j^2\gamma_j}{2\alpha_i^2\gamma_i} > 1 \Leftrightarrow \sqrt{\frac{\alpha_j^2\gamma_j}{2\alpha_i^2\gamma_i}} > 1$$

demonstrates that this case would require a discount factor for agent i which exceeds unity (see Table 4.1). Using this relationship and a similar computation for the optimal policies of agent i gives a general description of both agents' optimal transfer policies which are summarized by Tables 4.3 and 4.4.

Before turning to combining Tables 4.3 and 4.4, the general result of developing an agent's optimal policy should be summarized:

- The sign of transfers for an agent's optimal policy depends crucially on the indicator which describes the other agent's attribute. The other agent is mainly a contributor to a public good if maximization of both agents' utilities gives him/her positive bargaining gains before transfers; he/she is mainly a seller if maximization of both agents' utilities gives him/her negative bargaining gains before transfers.
- If the other agent is mainly a contributor to a public good, the optimal policies induce transfers from this agent to the optimizing agent or no transfers. If transfers are paid, the optimizing agent realizes an effort level which falls short of the level of an optimal enforceable contract whereas the other agent realizes the effort level which an optimal enforceable contract would specify for him/her (see (4.13)).
- If the other agent is mainly a seller, the optimal policies induce transfers from the optimizing agent to the other agent or no transfers. If transfers are paid, the optimizing agent realizes the effort level which an optimal enforceable contract would specify for him/her whereas the other agent's effort level falls short of this level (see (4.19)).

The last point is worth emphasizing. If an agent is a seller, the optimizing agent pays him/her but does not require that he/she realizes the Pareto-optimal effort level. The optimizing agent, however, chooses the Pareto-optimal level as the level which maximizes his/her bargaining gains. Note that this result is not counterintuitive when the levels are compared. The seller may have low costs and low benefits whereas the optimizing agent has high costs and high benefits. If the optimizing agent realizes his/her Pareto-optimal level, this level may still fall short of the other agent's (less than Pareto-optimal) level because high costs of the optimizing agent lead to a comparably low Pareto-optimal level.

Table 4.3 Optimal transfer policies for agent j

Range of the indicator describing agent i's attribute	Range of agent i's discount factor	Sign of optimal transfer policies
$\dfrac{2\alpha_i^2\gamma_i}{\alpha_j^2\gamma_j} \geq 1$	$\delta_i > \sqrt{\dfrac{\alpha_j^2\gamma_j}{2\alpha_i^2\gamma_i}}$	$T_j^* < 0$
(public good contributor attribute)	$\delta_i \leq \sqrt{\dfrac{\alpha_j^2\gamma_j}{2\alpha_i^2\gamma_i}}$	$T_j^* = 0$
$\dfrac{2\alpha_i^2\gamma_i}{\alpha_j^2\gamma_j} < 1$	$\delta_i > \dfrac{2\alpha_i^2\gamma_i}{\alpha_j^2\gamma_j}$	$T_j^* > 0$
(seller attribute)	$\delta_i \leq \dfrac{2\alpha_i^2\gamma_i}{\alpha_j^2\gamma_j}$	$T_j^* = 0$

So far, the discussion has demonstrated that transfers have already increased the complexity of optimal policies. The optimal policies of two agents define the relevant range of the concession lines. Tables 4.3 and 4.4 can be combined in order to outline all possibilities. Comparing the second columns of each table reveals that both agents cannot qualify as a seller because $(2\alpha_i^2\gamma_i)/(\alpha_j^2\gamma_j) < 1$ and $(2\alpha_j^2\gamma_j)/(\alpha_i^2\gamma_i) < 1$ contradict each other. Table 4.5 collects all possible twelve combinations. These combinations can be subdivided into three sets (indicated by A, B and C) which describe the different combinations of attributes. In each of these sets, four elements give the possible combinations of zero and non-zero optimal transfers (indicated by numbers 1 to 4) which depend on the discount factors. The next section will use this table to discuss the different concession lines after it has determined concession policies.

Table 4.4 Optimal transfer policies for agent i

Range of the indicator describing agent i's attribute	Range of agent j's discount factor	Sign of optimal transfer policies
$\dfrac{2\alpha_j^2\gamma_j}{\alpha_i^2\gamma_i} \geq 1$	$\delta_j > \sqrt{\dfrac{\alpha_i^2\gamma_i}{2\alpha_j^2\gamma_j}}$	$T_i^* < 0$
(public good contributor attribute)	$\delta_j \leq \sqrt{\dfrac{\alpha_i^2\gamma_i}{2\alpha_j^2\gamma_j}}$	$T_i^* = 0$
$\dfrac{2\alpha_j^2\gamma_j}{\alpha_i^2\gamma_i} < 1$	$\delta_j > \dfrac{2\alpha_j^2\gamma_j}{\alpha_i^2\gamma_i}$	$T_i^* > 0$
(seller attribute)	$\delta_j \leq \dfrac{2\alpha_j^2\gamma_j}{\alpha_i^2\gamma_i}$	$T_i^* = 0$

Table 4.5 *Optimal transfer policies for both agents*

Agent i's gains from full cooperation... Agent j's gains from full cooperation...	are positive $\dfrac{2\alpha_i^2\gamma_i}{\alpha_j^2\gamma_j} \geq 1$ (public good contributor attribute)	are negative $\dfrac{2\alpha_i^2\gamma_i}{\alpha_j^2\gamma_j} < 1$ (seller attribute)
	A: $T_j^* \leq 0 \wedge T_i^* \geq 0$ (both receivers)	B: $T_j^* \geq 0 \wedge T_i^* \geq 0$ (donor/receiver)
are positive $\left(2\alpha_i^2\gamma_i \big/ \alpha_j^2\gamma_j\right) \geq 1$ (public good contributor attribute)	A$_1$: $T_j^* < 0 \wedge T_i^* > 0$	B$_1$: $T_j^* > 0 \wedge T_i^* > 0$
	A$_2$: $T_j^* = 0 \wedge T_i^* > 0$	B$_2$: $T_j^* = 0 \wedge T_i^* > 0$
	A$_3$: $T_j^* < 0 \wedge T_i^* = 0$	B$_3$: $T_j^* > 0 \wedge T_i^* = 0$
	A$_4$: $T_j^* = 0 \wedge T_i^* = 0$	B$_4$: $T_j^* = 0 \wedge T_i^* = 0$
are negative $\left(2\alpha_i^2\gamma_i \big/ 2\alpha_i^2\gamma_i\right) < 1$ (seller attribute)	C: $T_j^* \leq 0 \wedge T_i^* \leq 0$ (receiver/donor)	
	C$_1$: $T_j^* < 0 \wedge T_i^* < 0$	
	C$_2$: $T_j^* = 0 \wedge T_i^* < 0$	
	C$_3$: $T_j^* < 0 \wedge T_i^* = 0$	
	C$_4$: $T_j^* = 0 \wedge T_i^* = 0$	

4.4 CONCESSION POLICIES AND CONCESSION LINES

In Chapter 3, developing the concession lines was straightforward because both agents' effort policy determines the compliance constraints and the contract curve directly. When transfers are allowed for, two complications arise for determining the relevant concession line. First, Table 4.5 has indicated that twelve possible combinations of optimal policies exist. Consequently, there are not only three candidates (both compliance constraints and the contract curve or a single point if $\delta_i \, \delta_j < 0.25$) as in Chapter 3 but twelve concession lines are possible which depend on the agents' attributes and their discount factors. Second, introducing transfers enlarges the set of possible concessions of an agent. For example, agent i may concede by proposing a lower effort level for agent j, a higher effort level for him-/herself and/or a lower transfer level.

This section begins with developing concession policies. It takes agent j's optimal policy as the starting-point and determines the role of efforts and transfers for concessions away from agent j's optimal policy. Several cases which are indicated by Table 4.5 will be distinguished and they will be shown to have different impacts on the concession line. The concession line will be developed by employing a strategic bargaining parameter. This section imposes no feasibility restrictions on the concession line which will be discussed in the next section.

Since the analysis of all relevant cases is involved, readers not interested in an explicit discussion of concession policies may restrict their attention to Tables 4.6 and 4.7 on pages 159 and 167, respectively, and to formulas (4.36) and (4.46) on pages 154 and 158, respectively. This section demonstrates that only two or one out of three concession instruments (the two effort levels and the level of transfers) are changed. Tables 4.6 and 4.7 summarize the concession policies of agent j when agent j makes concessions away from his optimal policy. They show that either the effort levels or one effort level and the transfer level are changed until the Pareto-optimal effort levels are reached. Then, concessions will be made exclusively by changing the transfer level. Table 4.7 demonstrates in particular that transfers may not play a role for concession policies over a wide range because the concession line only alters both effort levels and leaves transfers on their zero level.

Marginal Bargaining Gains along the Concession Line

Before developing the concession line for the cases at hand, a reformulation of the definition which determines the concession line in general will be helpful for this model. The concession line may be defined by the dual condition that every point maximizes one agent's bargaining gains for a given non-negative level of the other agent's bargaining gains. These bargaining gains (of the other agent) must lie in the range between the other agent's maximum bargaining gains and the bargaining gains which he/she realizes if one agent's bargaining gains are maximized.

Let (U_i', U_j') and (U_i'', U_j'') denote any two vectors of bargaining gains which belong to the set of efficient agreements. Defining

$$\Delta U_i := U_i' - U_i'' \quad \Delta U_j := U_j' - U_j'',$$

allows concessions along the concession line to be discussed. ΔU_i (ΔU_j) indicates a change of agent i's (j's) bargaining gains from one point (denoted by a double prime) to another point (denoted by a prime) on the concession line. The assumption that (U_i', U_j') and (U_i'', U_j'') belong to the set of efficient agreements implies that the change of bargaining gains, ΔU_i, must be maximized for a given change of the other agent's bargaining gains, ΔU_j. This means that the increase (decrease) in one agent's bargaining gains must be maximized (minimized) if the other agent's bargaining gains are decreased (increased). (Note that $\Delta U_i \Delta U_j < 0$ holds on the concession line.) The intuition behind ΔU_i and ΔU_j is clear: if an agent's bargaining gains are changed, the other agent's bargaining gains should be maximized on the concession line (this condition holds in general independent of the specific model, see also (2.45) and (2.46)). For example, if agent i's bargaining gains are increased by a given amount, the decrease in agent j's bargaining gains should be minimized, that is, his/her negative increase should be maximized.

For marginal concessions which determine the strategic bargaining solution, the total differentials of both agents' utility function are given by

$$dU_i = \frac{\partial U_i}{\partial E_i} dE_i + \frac{\partial U_j}{\partial E_j} dE_j + \frac{\partial U_i}{\partial T} dT = \left[\alpha_i - \gamma_i E_i\right] dE_i + \alpha_i dE_j + dT,$$

$$dU_j = \frac{\partial U_j}{\partial E_i} dE_i + \frac{\partial U_j}{\partial E_j} dE_j + \frac{\partial U_j}{\partial T} dT = \alpha_j dE_i + \left[\alpha_j - \gamma_j E_j\right] dE_j - dT.$$

$$(4.23)$$

Along the concession line dU_i (dU_j) must be maximized for any dU_j (dU_i). This is the consequence of moves along the set of efficient agreements (see for example, Figure 2.3). If dU_i (dU_j) were not maximized for a certain dU_j (dU_i), a marginal move away from a point on the payoff frontier would lead to a point below the payoff frontier.

Concession Policies for a Negative Optimal Transfer Level

The first case of concession policies assumes that T_j^* holds and that agent j considers concession policies away from his/her optimal policy. Optimal policy for agent j means transfer and effort levels according to (4.13). The term concession policies can be given some further clarification now. Suppose that agent j has not the whole but 'almost' the whole bargaining power in the bargaining process. This enables him/her to specify an agreement close to his/her optimal policies. Taking the optimal policy result as a starting-point, agent j considers how to concede a 'little bit' in order to match the strategic bargaining solution, that is, he/she intends to minimize his/her loss compared to his/her optimal policy.

For the problem here, the relevant compliance constraint is given by the third line of (4.8). This section will not consider the general case of concessions which changes all three variables. Instead, it will focus on two out of three variables which are changed. This restriction is justified if all possible candidates can be ranked by the Pareto criterion. In this case, concession policies employ only two specific variables. This section will show that such a Pareto dominance holds for a range which starts at an agent's optimal policy. Three cases for potential concession policies can be distinguished and interpreted by Figure 4.3.

- Concessions along the compliance constraint curve: This concession policy should indicate that concessions are made according to the old policy which suppressed transfers. The notion 'along the compliance constraint curve' originates from Figure 4.3 which depicts this policy as moves along the line without changing the transfer level.
- Concession by an upward shift of the compliance constraint curve: This concession policy does not change agent i's efforts but varies transfers and agent j's efforts. As this concession policy will be shown to imply an increase in agent j's efforts, it is called an upward shift of the compliance constraint curve.
- Concessions by a rightward shift of the compliance constraint curve: This concession policy does not change agent j's efforts but changes transfers and agent i's efforts. As this concession policy will be shown to imply an increase in agent i's efforts, it is called a rightward shift of the compliance constraint curve.

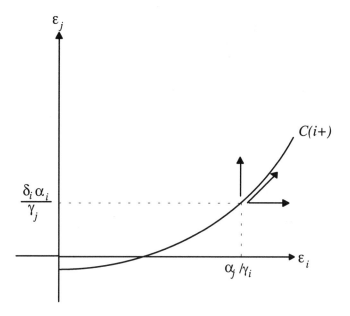

Figure 4.3   Different concession policies for $T_j^* < 0$

All cases are indicated by arrows in Figure 4.3. The formal computation starts with concessions along the compliance constraint curve. Differentiating the third line of (4.8) for $dT = 0$ gives

$$\left.\frac{dE_i}{dE_j}\right|_{dT=0} = \frac{\delta_i \alpha_i}{\gamma_i \varepsilon_i}. \tag{4.24}$$

Equation (4.24) can be inserted into the total differentials (4.23). The subscript $dT = 0$ denotes that only marginal concessions along the compliance constraint are considered:

$$dU_i|_{dT=0} = \alpha_i(1-\delta_i)dE_j,$$

$$dU_j\Big|_{dT=0} = \left[\frac{\delta_i \alpha_i \alpha_j}{\gamma_i \varepsilon_i} + \alpha_j - \gamma_j E_j\right]dE_j. \tag{4.25}$$

The subscript $dE_i = 0$ denotes concessions by an upward shift of the compliance constraint curve. Differentiating the third line of (4.8) for $dE_i = 0$ gives

$$\frac{dT}{dE_j}\bigg|_{dE_i=0} = -\delta_i\alpha_i \tag{4.26}$$

and produces total differentials:

$$dU_i\big|_{dE_i=0} = \alpha_i\left(1-\delta_i\right)dE_j,$$

$$dU_j\big|_{dE_i=0} = \left(\alpha_j - \gamma_j E_j + \delta_i\alpha_i\right)dE_j. \tag{4.27}$$

Finally, the subscript $dE_j = 0$ denotes concessions by a rightward shift of the compliance constraint curve. Differentiating the third line of (4.8) for $dE_j = 0$ gives

$$\frac{dT}{dE_i}\bigg|_{dE_j=0} = \gamma_j\varepsilon_i \tag{4.28}$$

and produces total differentials:

$$dU_i\big|_{dE_j=0} = 0,$$

$$dU_j\big|_{dE_j=0} = \left(\alpha_j - \gamma_i\varepsilon_i\right)dE_i. \tag{4.29}$$

Equations (4.25), (4.27) and (4.29) describe marginal moves which hold one variable constant and set the compliance constraint defined by agent i equal to zero.

Equations (4.27) and (4.29) explain why the shifts are called upwards and rightwards, respectively. Equation (4.27) shows that an improvement in agent i's utility, that is, $dU_i > 0$, is only possible if agent j increases his/her efforts, that is, $dE_j > 0$, for an unchanged effort level of agent i. Thus, increasing E_j for this kind of policy justifies the notion of an upward shift as is indicated in Figure 4.3. The same line of reasoning applies in explaining the notion of a rightward shift in (4.29) because dU_j is negative only for a positive dE_i when changes away from the optimal policies are considered. Equation (4.29), however, directly demonstrates that concessions by a rightward shift of the compliance constraint curve are useless: the utility of agent i cannot be improved upon if agent j's efforts are kept constant. A policy which leaves one agent's utility level constant but weakens the utility level of the other agent must be ruled out as a candidate for concession policies.

By comparing (4.25) and (4.27), the relevant policies can be determined. Comparability is possible because all terms in (4.25) and (4.27) are given in changes of agent j's efforts. Comparing dU_i in (4.25) and (4.27) reveals that agent i is indifferent between concessions along the compliance constraint curve and concessions by an upward shift of the compliance constraint curve. Agent j, however, weakly prefers the upward-shift concession. If agent j introduces a concession policy along the compliance constraint curve, (4.25) indicates that agent i's efforts will be increased as well. Increasing agent i's efforts means that this effort level exceeds the Pareto-optimal level which (4.13) has specified for the optimal policy of agent j:

$$E_i \geq \frac{\alpha_i + \alpha_j}{\gamma_i} \Rightarrow dU_j\bigg|_{dT=0} \leq \left(\delta_i\alpha_i + \alpha_j - \gamma_j E_j\right)dE_j. \quad (4.30)$$

Equation (4.30) uses this line of reasoning to determine the lowest level of utility losses implied by concessions along the compliance constraint curve. If agent j introduces a concession policy which shifts the compliance constraint curve upwards, the utility loss is given by the second line of (4.27). Which policy will be pursued by agent j? (Recall that agent i is indifferent as to which policy is followed.) Preference for the upward-shift policy can easily be proved by starting with changes away from agent j's optimal policy (4.13). For $E_j = (\delta_i\alpha_i + \alpha_j)/\gamma_j$, the second lines of (4.25) and (4.27) coincide. However, for all other E_j marginally exceeding $(\delta_i\alpha_i + \alpha_j)/\gamma_j$, the upward-shift policy produces lower marginal utility losses than does the other policy. Equation (4.30) shows that the second line of (4.25) exceeds the second line of (4.27) in absolute terms. Except for the optimal policy, agent j strongly prefers the upward-shift policy, which induces lower marginal losses than does the other policy.

These results give a straightforward interpretation for the concession line. Starting from the optimal policy, the upward-shift concessions maximize agent j's bargaining gains for a given marginal increase in agent i's bargaining gains. As this result holds unambiguously in the close neighbourhood of agent j's optimal policy and agent i is indifferent, every rational concession policy starts with (4.27).

Note that this conclusion is a general one and not due to the restriction to two policy instruments. If agent i's marginal utility changes coincide for concessions along the compliance constraint curve and upward-shift concessions, they also coincide for any combination of both policies. But if upward-shift concessions benefit agent j more than any other policy, a rational concession policy will be restricted to this option. Any other policy left the bargaining gains of agent j unexploited without changing agent i's increase in bargaining gains.

If the first part of the concession line is defined by shifting the compliance constraint curve upwards, agent j's efforts are increased whereas the transfers are decreased (see (4.26)). This result was based on the fact that agent i's compliance constraint defines a part of the concession line in the close neighbourhood of agent j's optimal policy. This conclusion is similar to the model of Chapter 3 which demonstrated that the agents' compliance constraint curves define the concession line until the contract curve is reached. This model, however, does not employ a contract curve as a set of (in the usual sense) Pareto-optimal effort vectors but only one vector of Pareto-optimal effort levels which has been developed in Section 4.2. If $E_i = (\alpha_i + \alpha_j)/\gamma_i$ and $E_j = (\alpha_i + \alpha_j)/\gamma_j$, the total bargaining gains are maximized. If the concession line reaches this point, compliance constraints do not bind because transfers are allowed to play their role in distributing the maximum bargaining gains.

This conclusion has a straightforward implication which can be developed by considering upward-shift concessions. Conceding means that agent j will increase his/her own efforts and decrease transfers. If the Pareto-optimal point is reached, transfers have been decreased to

$$T = \frac{\alpha_j^2}{2\gamma_i} - \frac{\delta_i \alpha_i^2}{\gamma_j} \tag{4.31}$$

(compare with the optimal transfers in (4.13)). The marginal utility changes are equalized at this point:

$$dU_j \Big|_{dE_i=0} \left(E_i = \frac{\alpha_i + \alpha_j}{\gamma_i}, E_j = \frac{\alpha_i + \alpha_j}{\gamma_j} \right) = -\alpha_i (1 - \delta_i) dE_j$$

$$= -dU_i \Big|_{dE_i=0} \left(E_i = \frac{\alpha_i + \alpha_j}{\gamma_i}, E_j = \frac{\alpha_i + \alpha_j}{\gamma_j} \right)$$

$$\Rightarrow \frac{dU_j \big|_{dE_i=0}}{dU_i \big|_{dE_i=0}} \left(E_i = \frac{\alpha_i + \alpha_j}{\gamma_i}, E_j = \frac{\alpha_i + \alpha_j}{\gamma_j} \right) = -1 \tag{4.32}$$

Note that dU_j is lower for all E_j which fall short of the Pareto-optimal level. The second line of (4.32) gives the minimum of agent j's marginal utility decreases for a given marginal utility increase of agent i. According to the upward-shift policy,

$$\frac{dU_j \big|_{dE_i=0}}{dU_i \big|_{dE_i=0}} \left(E_i = \frac{\alpha_i + \alpha_j}{\gamma_i}, E_j < \frac{\alpha_i + \alpha_j}{\gamma_j} \right) > -1$$

$$\frac{dU_j \big|_{dE_i=0}}{dU_i \big|_{dE_i=0}} \left(E_i = \frac{\alpha_i + \alpha_j}{\gamma_i}, E_j > \frac{\alpha_i + \alpha_j}{\gamma_j} \right) < -1 \qquad (4.33)$$

defines the behaviour of the marginal utility ratio in the neighbourhood of the Pareto-optimal level. Note that the necessary marginal utility losses, that is, $-dU_j$, increase as E_j increases, reach unity for the Pareto-optimal efforts, and exceed unity if agent j pursues this policy if E_j should surmount its Pareto-optimal level. In this case, the marginal losses of agent j must exceed the marginal gains of agent i in absolute terms. But agent j has the alternative option to concede only by increasing transfers without changing the Pareto-optimal effort levels:

$$dU_i \big|_{dE_i=dE_j=0} = 1, \quad dU_j \big|_{dE_i=dE_j=0} = -1,$$

$$\frac{dU_j \big|_{dE_i=dE_j=0}}{dU_i \big|_{dE_i=dE_j=0}} = -1. \qquad (4.34)$$

Equation (4.34) demonstrates that an upward-shift policy beyond agent j's Pareto-optimal level would sacrifice unnecessarily high bargaining gains. This result is due to the maximized total bargaining gains at this effort level which imply that transfers are the only relevant concession instrument. From (4.33) and (4.34), it is clear that

- concessions away from agent j's optimal policy increase agent j's efforts and decrease transfers until the Pareto-optimal levels are reached, and
- concessions are made later by transfers without changing both agents' effort policy.

Note that transfers change their direction along the two parts of the concession line: they are decreased if agent i's compliance constraint defines the concession line; they are increased after the Pareto-optimal effort levels have been reached.

Upward-shift concessions can also be described by a bargaining parameter similar to the bargaining parameter employed in Chapter 3. The effort level of agent j can be described by D_{in} with

$$E_j = \frac{D_{in}\alpha_i + \alpha_j}{\gamma_j}, \quad \delta_i \leq D_{in} \leq 1. \tag{4.35}$$

D_{in} denotes the bargaining parameter for the compliance constraint defined by agent i and for negative transfers. According to (4.35), the bargaining gains of both agents are given by

$$U_i = (1-\delta_i)\frac{D_{in}\alpha_i^2}{\gamma_j}, \quad U_j = \frac{\alpha_j^2}{2\gamma_i} + (2\delta_i - D_{in})\frac{D_{in}\alpha_i^2}{2\gamma_j}. \tag{4.36}$$

Until now, this chapter has not yet discussed the feasibility of the Pareto-optimal effort levels but has assumed implicitly that the compliance constraint defined by i from j's optimal policy to the Pareto-optimal levels and some exclusively transfer-based concessions are feasible. The section will also keep this implicit assumption for the following computation which assumes that $T_j^* > 0$ holds. This optimistic assumption will be dropped in the next section, which discusses feasibility aspects and the behaviour of the concession line in general.

Concession Policies for a Positive Optimal Transfer Level

For $T_j^* > 0$, three cases can also be distinguished which are depicted in Figure 4.4:

- Concessions along the compliance constraint curve: These concessions leave the transfer level unchanged. Differentiating the first line of (4.8) for $dT = 0$ gives

$$\left.\frac{dE_j}{dE_i}\right|_{dT=0} = \frac{\gamma_i \varepsilon_i}{\delta_i \alpha_i} \tag{4.37}$$

Inserting (4.37) into the total differentials implies

$$dU_i\big|_{dT=0} = \frac{1-\delta_i}{\delta_i}\gamma_i\varepsilon_i dE_i,$$

$$dU_j\big|_{dT=0} = \left[\alpha_j - \frac{\gamma_i\gamma_j}{\delta_i\alpha_i}\varepsilon_i\varepsilon_j\right]dE_i. \tag{4.38}$$

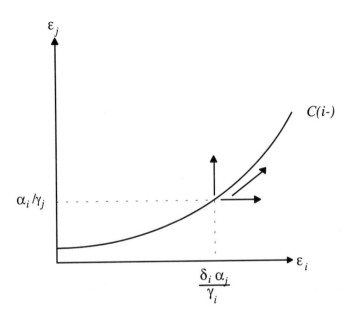

Figure 4.4   Different concession policies for $T_j^* > 0$

- Concessions by a rightward shift of the compliance constraint curve: These concessions leave agent *j*'s effort level unchanged. Differentiating the first line of (4.8) for $dE_j = 0$ gives

$$\frac{dT}{dE_i}\bigg|_{dE_j=0} = \frac{\gamma_i}{\delta_i}\varepsilon_i. \tag{4.39}$$

Inserting (4.43) into the total differentials implies

$$dU_i\Big|_{dE_j=0} = \frac{1-\delta_i}{\delta_i}\gamma_i\varepsilon_i dE_i,$$

$$dU_j\Big|_{dE_j=0} = \left[\alpha_j - \frac{\gamma_i}{\delta_i}\varepsilon_i\right]dE_i. \qquad (4.40)$$

- Concessions by an upward shift of the compliance constraint curve: These concessions leave agent i's effort level unchanged but any policy based on this policy is of no use. This result can easily be demonstrated by differentiating U_i totally:

$$dU_i = \frac{1-\delta_i}{\delta_i}\gamma_i\varepsilon_i dE_i$$

The total differential without any restriction to variations shows that the upper lines in (4.38) and (4.40) are not identical by chance because the total differential depends only on dE_i. If dE_i is set zero, agent i's bargaining gains cannot be improved upon. (A similar argument has explained the inferiority of rightward shifts for $T_j^* < 0$).

Accordingly, the analysis also can be restricted to two cases. Comparing (4.38) and (4.40) shows that agent i is indifferent between a rightward-shift policy and a policy along the compliance constraint curve. A rightward-shift policy increases the transfers and increases agent i's efforts; a policy along the compliance constraint curve increases both agents' efforts. Contrary to agent i's indifference, agent j weakly prefers the rightward-shift concession. Equation (4.19) has demonstrated that agent j's optimal policy requires that he/she realizes the Pareto-optimal level. Concessions along the compliance constraint curve imply

$$E_j \geq \frac{\alpha_i + \alpha_j}{\gamma_j} \Rightarrow dU_j\Big|_{dT=0} \leq \left[\alpha_j - \frac{\gamma_i}{\delta_i}\varepsilon_i\right]dE_i \qquad (4.41)$$

whereas the rightward-shift policy implies a marginal utility change which is given by the second line of (4.40). As the arguments are similar to those explicitly developed for the $T_j^* < 0$ case, the line of reasoning is obvious: agent j prefers the rightward-shift policy because the corresponding marginal utility losses fall short of the corresponding marginal utility losses which concessions along the compliance constraint curve incurred. Consequently, a rational concession policy will be restricted to a rightward-shift policy when agent j makes concessions away from his/her optimal policy. The same arguments hold which were given for the $T_j^* < 0$ case and also carry over to concession policies when the Pareto-optimal levels are reached. If the Pareto-optimal levels are reached, transfers have been increased to

$$T = \frac{1}{\delta_i} \frac{\alpha_j^2}{2\gamma_i} - \frac{\alpha_i^2}{\gamma_j} \tag{4.42}$$

(compare with the optimal transfers in (4.19)). The marginal utility changes are equalized at this point:

$$dU_j\Big|_{dE_j=0} \left(E_i = \frac{\alpha_i + \alpha_j}{\gamma_i}, \ E_j = \frac{\alpha_i + \alpha_j}{\gamma_j} \right) = -\alpha_j \frac{1-\delta_i}{\delta_i} dE_i =$$

$$-dU_i\Big|_{dE_j=0} \left(E_i = \frac{\alpha_i + \alpha_j}{\gamma_i}, \ E_j \frac{\alpha_i + \alpha_j}{\gamma_j} \right)$$

$$\Rightarrow \frac{dU_j\big|_{dE_j=0}}{dU_i\big|_{dE_j=0}} \left(E_i = \frac{\alpha_i + \alpha_j}{\gamma_i}, E_j = \frac{\alpha_i + \alpha_j}{\gamma_j} \right) = -1, \tag{4.43}$$

and the marginal utility change ratios are

$$\frac{dU_j\big|_{dE_j=0}}{dU_i\big|_{dE_j=0}} \left(E_i < \frac{\alpha_i + \alpha_j}{\gamma_i}, \ E_j = \frac{\alpha_i + \alpha_j}{\gamma_j} \right) > -1,$$

$$\frac{dU_j\big|_{dE_j=0}}{dU_i\big|_{dE_j=0}} \left(E_i > \frac{\alpha_i + \alpha_j}{\gamma_i}, \ E_j = \frac{\alpha_i + \alpha_j}{\gamma_j} \right) < -1. \tag{4.44}$$

Note that dU_i is not fixed but depends on agent i's effort level in this case. If the concession policy which is based on rightward shifts reaches the Pareto-optimal levels, the marginal losses of agent j should exceed the marginal gains of agent i. But agent j also has the option to concede by only increasing the transfer level without changing the Pareto-optimal effort levels. Equation (4.34) also holds for this case and induces a change of concession policies when E_i reaches its Pareto-optimal level. Summarizing, (4.34) and (4.44) imply that

- concessions away from agent j's optimal policy increase agent i's efforts and the transfer level until the Pareto-optimal levels are reached, and
- concessions are made later by transfers without changing either agent's effort policy.

Rightward-shift concessions can also be described by a bargaining parameter. The effort level of agent i can be given by D_{ip} with

$$E_i = \frac{\alpha_i + D_{ip}\alpha_j}{\gamma_i}, \quad \delta_i \le D_{ip} \le 1. \tag{4.45}$$

D_{ip} denotes the bargaining parameter for the compliance constraint defined by agent i and for positive transfers. According to (4.45), the bargaining gains of both agents are given by

$$U_i = \frac{D_{ip}^2\alpha_j^2}{2\gamma_i}\frac{1-\delta_i}{\delta_i},$$

$$U_j = \frac{\alpha_i^2}{2\gamma_j} + \left[1 - \frac{D_{ip}}{2\delta_i}\right]\frac{D_{ip}\alpha_j^2}{\gamma_i}. \tag{4.46}$$

Both cases which have been developed in this section are indicated by A_1, B_1 and C_1 in Table 4.5 because developing D_{jn} and D_{jp} is a similar process. Table 4.6 summarizes both cases for moves away from agent j's optimal policy when the optimal policy implies non-zero transfers.

Note that the sign of dT changes when the Pareto-optimal levels are reached if agent j's optimal policy specifies that transfers will be received from agent i, that is, $T_j^* < 0$. On the contrary, the sign of dT is positive for both concession policies if agent j's optimal policy specifies payment to agent i. In this case, moving away from the optimal policy means increasing the transfers in exchange for an increase in agent i's efforts until the Pareto-optimal levels are reached. For $T_j^* < 0$, moving away from the optimal policy means decreasing the transfers, that is, increasing the transfers which agent j receives, in exchange for an increase in agent j's efforts until the Pareto-optimal levels are reached. In both cases, transfers compensate for an increased effort level.

Table 4.6 Concession policies for $T_j^* \neq 0$

		$T_j^* < 0$	$T_j^* > 0$
Concession policies of agent j	until the Pareto-optimal levels are reached	$dE_j > 0, dT < 0,$ $dE_i = 0$ according to $\dfrac{dT}{dE_j} = -\delta_i \alpha_i$	$dE_i > 0, dT > 0,$ $dE_j = 0$ according to $\dfrac{dT}{dE_i} = \dfrac{\gamma_i}{\delta_i}\varepsilon_i$
	beyond the Pareto-optimal levels	$dT > 0, dE_i = 0, dE_j = 0$	

Concession Policies for a Zero Optimal Transfer Level

Dealing with concession policies if the optimal policies of agent j induce T_j^* $= 0$ requires that two different cases be distinguished (see Table 4.3):

- Case I:

$$\frac{2\alpha_i^2 \gamma_i}{\alpha_j^2 \gamma_j} \geq 1 \wedge \delta_i \leq \sqrt{\frac{\alpha_j^2 \gamma_j}{2\alpha_i^2 \gamma_i}}$$

(see cases A_2, A_4, C_2 and C_4 in Table 4.5).
- Case II:

$$\frac{2\alpha_i^2 \gamma_i}{\alpha_j^2 \gamma_j} < 1 \wedge \delta_i \leq \frac{2\alpha_i^2 \gamma_i}{\alpha_j^2 \gamma_j}$$

(see cases B_2 and B_4 in Table 4.5).

Case I deals with zero transfers when agent i qualifies as a public good contributor; Case II deals with zero transfers when agent i qualifies as a seller. Discussing both cases will complete the section on determining concession policies because the two cases for non-zero transfers, the two cases i zero transfers and their analogies for agent i define all conceivable cases given in Table 4.5. After developing concession policies for both cases in this section, the feasibility of concession lines for the different cases of Table 4.5 will be dealt with in the following section.

In Case I, the optimal policy for agent j is given either by (3.23) or (3.33) and (3.34). Equations (3.33) and (3.34) assume that both agents' discount factors are extremely low and make both agents' optimal policies coincide. As neither agent is able to improve on his/her utilities, the strategic bargaining solution is obvious and no concession policy is possible if $T_i^* = T_j^* = 0$ holds. One could imagine that transfers could play their role if the optimal policies of one agent implied a non-zero transfer level. However, a non-zero optimal transfer level and the assumption δ_i, $\delta_j < 0.25$ which implies (3.33) and (3.34) give a contradiction. Assume that agent j's optimal policy would require a non-zero transfer level. According to Tables 4.3 and 4.4,

$$\delta_i < 0.25, \delta_j < 0.25, \quad \frac{2\alpha_i^2\gamma_i}{\alpha_j^2\gamma_j} \ge 1, \quad \frac{2\alpha_j^2\gamma_j}{\alpha_i^2\gamma_i} \ge 1,$$

$$\delta_i > \sqrt{\frac{\alpha_j^2\gamma_j}{2\alpha_i^2\gamma_i}}, \quad \delta_j \le \sqrt{\frac{\alpha_i^2\gamma_i}{2\alpha_j^2\gamma_j}}, \quad \Rightarrow \frac{1}{4} > \sqrt{\frac{\alpha_j^2\gamma_j}{2\alpha_i^2\gamma_i}} \Leftrightarrow \frac{2\alpha_j^2\gamma_j}{\alpha_i^2\gamma_i} < \frac{1}{4}$$

$$(4.47)$$

should hold but (4.47) leads to a contradiction because the last line contradicts the attribute assumption for agent i. Alternatively, assume that agent i's optimal policy implied a non-zero transfer level. According to Tables 4.3 and 4.4,

$$\delta_i < 0.25, \delta_j < 0.25, \quad \frac{2\alpha_i^2\gamma_i}{\alpha_j^2\gamma_j} \ge 1, \quad \frac{2\alpha_j^2\gamma_j}{\alpha_i^2\gamma_i} \ge 1,$$

$$\delta_i \le \sqrt{\frac{\alpha_j^2\gamma_j}{2\alpha_i^2\gamma_i}}, \quad \delta_j > \sqrt{\frac{\alpha_i^2\gamma_i}{2\alpha_j^2\gamma_j}}, \quad \Rightarrow \frac{1}{4} > \sqrt{\frac{\alpha_i^2\gamma_i}{2\alpha_j^2\gamma_j}} \Leftrightarrow \frac{2\alpha_i^2\gamma_i}{\alpha_j^2\gamma_j} < 1$$

$$(4.48)$$

should hold but (4.48) contains a similar contradiction as (4.47). Thus, if δ_i, $\delta_j < 0.25$ is valid, transfers cannot cure or mitigate the compliance problem but the bargaining solution is determined by (3.33) and (3.34) even if transfers are allowed.

The remaining discussion of Case I assumes that the concession policies to be developed are not restricted because certain policies are not feasible. When discussing the cases indicated by Table 4.5 in Section 4.5, feasibility will be dealt with explicitly. Thus, the discussion develops concession policies for the case that these policies are allowed by both agents' compliance constraints. If δ_i, $\delta_j > 0.25$ holds, the optimal policy for agent j is given by (3.23). Starting from these effort levels, three cases for concession policies can be distinguished again:

- concessions along the compliance constraint curve (*Ia*),
- concessions by an upward shift (*Ib*) and
- concessions by a rightward shift.

The last case was proved to be of no use by (4.29). Note that (4.25), (4.27) and (4.29) still hold because they describe marginal changes in general for a non-positive transfer level. Therefore, it is obvious that concession policies by a rightward shift are also of no use for the zero-transfer case and can be excluded from the set of policy candidates. The relevant cases are denoted by *Ia* and *Ib*.

The superiority of the upward-shift policy for the non-zero transfer case was based on the result that agent j's optimal policy requires that agent i realizes his/her Pareto-optimal effort level. This result does not hold for Case I:

$$
\frac{\alpha_i + \alpha_j}{\gamma_i} = \left[\frac{\alpha_j^2 \gamma_j}{2\alpha_i^2 \gamma_i} \frac{2\alpha_i^2 \alpha_j}{\gamma_i^2 \gamma_j} \right]^{1/3} + \frac{\alpha_j}{\gamma_i} \geq \left[\delta_i^2 \frac{2\alpha_i^2 \alpha_j}{\gamma_i^2 \gamma_j} \right]^{1/3} + \frac{\alpha_i}{\gamma_i} \tag{4.49}
$$

because

$$
\delta_i \leq \sqrt{\frac{\alpha_j^2 \gamma_j}{2\alpha_i^2 \gamma_i}} \Leftrightarrow \frac{\alpha_j^2 \gamma_j}{2\alpha_i^2 \gamma_i} \geq \delta_i^2 .
$$

Condition (4.49) shows that the range of agent i's discount factors which implies zero transfers for the optimal policy of agent j induces an effort level from agent i which does not exceed the Pareto-optimal one (see (3.23) and

Table 4.3). Agent i is indifferent between options Ia and Ib because his/her marginal gains are identical. For zero transfers defining his/her optimal policy, however, agent j prefers to make concessions away from his/her optimal policy, by conceding along the compliance constraint curve.

Let dU_{jIa} and dU_{jIb} denote the marginal utility changes which both concession policies produce. Reformulating (3.41), the marginal utility change implied by Ia for marginal moves away from agent j's optimal policy gives:

$$dU_{jIa}\left(d_i = \delta_i^2\right) = \frac{\delta_i \alpha_i \alpha_j}{\gamma_i} \frac{1-\left(d_i/\delta_i^2\right)}{\varepsilon_i} dE_j = 0. \tag{4.50}$$

Equation (4.50) uses the results of Section 3.3 which employed the strategic bargaining parameter d_i to describe strategic bargaining solutions on agent i's compliance constraint. The marginal utility change which option Ib implies for marginal moves away from agent j's optimal policy changes agent j's efforts and the transfer level according to (4.26). Inserting agent j's effort level for his/her optimal policy (see (3.23)) into the lower line of (4.27) gives

$$dU_{jIb}\left[E_j = \frac{\gamma_i}{2\alpha_i}\left[\delta_i^2 \frac{2\alpha_i^2 \alpha_j}{\gamma_i^2 \gamma_j}\right]^{2/3} + \frac{\alpha_j}{\gamma_j}\right]$$

$$= \left[\delta_i \alpha_i - \frac{\gamma_i \gamma_j}{2\alpha_i}\left[\delta_i^2 \frac{2\alpha_i^2 \alpha_j}{\gamma_i^2 \gamma_j}\right]^{2/3}\right] dE_j \le 0 \tag{4.51}$$

because

$$\delta_i^2 \le \frac{\alpha_j^2 \gamma_j}{2\alpha_i^2 \gamma_i} \Leftrightarrow \delta_i^2 \frac{2\alpha_i^2 \alpha_i}{\gamma_i^2 \gamma_j} \le \frac{\alpha_j^3}{\gamma_i^3}$$

$$\Rightarrow \delta_i \alpha_i - \frac{\gamma_i \gamma_j}{2\alpha_i}\left[\delta_i^2 \frac{2\alpha_i^2 \alpha_j}{\gamma_i^2 \gamma_j}\right]^{2/3} \le \delta_i - \frac{\alpha_j^2}{2\alpha_i^2 \gamma_i} \le 0.$$

Equation (4.51) demonstrates that the marginal utility change is negative if agent i's discount factor is not equal to the limit $(2\alpha_i^2\gamma_i)/(\alpha_j^2\gamma_j)$. Thus, agent j is only indifferent as to which option is chosen if $\delta_i = (2\alpha_i^2\gamma_i)/(\alpha_j^2\gamma_j)$ holds. In all other cases, he/she prefers to make concessions away from his optimal policy by *Ia*.

Comparing dU_{jIa} and dU_{jIb} determines concession policies in general:

$$E_i \begin{matrix} \le \\ > \end{matrix} \frac{\alpha_i + \alpha_j}{\gamma_i} \Leftrightarrow dU_{jIa} \begin{matrix} \ge \\ < \end{matrix} dU_{jIb} \qquad (4.52)$$

because

$$\frac{\delta_i \alpha_i \alpha_j}{\gamma_i \varepsilon_i} + \alpha_j - \gamma_j E_j \begin{matrix} \ge \\ < \end{matrix} \alpha_j - \gamma_j E_j + \delta_i \alpha_i \Leftrightarrow E_i \begin{matrix} \le \\ > \end{matrix} \frac{\alpha_i + \alpha_j}{\gamma_i}.$$

In general, agent j favours concessions along the compliance constraint curve until agent i's Pareto-optimal level is reached. When this level is reached, agent j favours upward-shift concessions. Both policies produce the same marginal changes for agent i. Note that this assertion is based on the assumption that $E_i = (\alpha_i + \alpha_j)/\gamma_i$ is feasible. If $\delta_i \ge (2\alpha_i^2\gamma_i)/(\alpha_j^2\gamma_j)$ holds, concession policies directly shift the compliance constraint curve upwards because the optimal policies for j require that agent i realizes his/her Pareto-optimal level.

In case II, agent i qualifies as a seller of services. The optimal policies for agent j are given either by (3.23) or by (3.33) and (3.34). As for Case I, it must be checked whether transfers can play a role if $\delta_i, \delta_j < 0.25$ holds. Assume first that agent j's optimal policy implies a non-zero transfer level. According to Tables 4.3 and 4.4, (4.53) must hold.

$$\delta_i < 0.25, \delta_i < 0.25, \frac{2\alpha_i^2\gamma_i}{\alpha_j^2\gamma_j} < 1, \frac{2\alpha_j^2\gamma_j}{\alpha_i^2\gamma_i} \ge 1,$$

$$\delta_i > \frac{2\alpha_i^2\gamma_i}{\alpha_j^2\gamma_j}, \delta_i \le \sqrt{\frac{\alpha_i^2\gamma_i}{2\alpha_j^2\gamma_j}} \qquad (4.53)$$

In contrast to Case I, (4.53) does not contain any contradiction. This means that $\delta_i, \delta_j < 0.25$ in Case II does not necessarily imply a strategic bargaining solution given by (3.33) and (3.34). If T_j^* is non-zero, the concession policies will follow the lines given above for this case, possibly limited by feasibility restrictions. If agent i's optimal policy implied a non-zero transfer level

$$\delta_i < 0.25, \delta_j < 0.25, \ \frac{2\alpha_i^2 \gamma_i}{\alpha_j^2 \gamma_j} < 1, \ \frac{2\alpha_j^2 \gamma_j}{\alpha_i^2 \gamma_i} \geq 1,$$

$$\delta_i \leq \frac{2\alpha_i^2 \gamma_i}{\alpha_j^2 \gamma_j}, \ \delta_i > \sqrt{\frac{\alpha_i^2 \gamma_i}{2\alpha_j^2 \gamma_j}} \qquad (4.54)$$

should hold which does not contain any contradiction either. If T_i^* is non-zero, the concession policies will follow the lines given above. Thus, if an agent qualifies as a seller, the result for δ_i, $\delta_j < 0.25$ is not fixed by (3.33) and (3.34) but transfers and bargaining may be possible. Numerical examples can easily demonstrate that the differences in benefits and costs between the agents must be large to meet (4.53) and (4.54). If efficiency gains, indicated by high benefit and cost differences, are large, Case II does not automatically determine zero optimal transfer levels for δ_i, $\delta_j < 0.25$ as Case I does, and discount factor combinations δ_i, $\delta_j < 0.25$ may imply zero or non-zero transfer policies. If they imply non-zero transfers, concession policies follow the lines given in Table 4.6. If they imply zero transfers, the bargaining solution is defined by (3.33) and (3.34) and cannot be improved upon.

The rest of the discussion of Case II assumes that optimal policies for agent j are given by (3.23) and that the concession policies to be developed are not restricted because certain policies are infeasible. This section does not consider the case $\delta_i < 0.25$, $\delta_j \geq 0.25$ or $\delta_i \geq 0.25$, $\delta_j < 0.25$ (see Figure 3.3) which combines the arguments given for δ_i, $\delta_j < 0.25$ and δ_i, $\delta_j \geq 0.25$. If δ_i, $\delta_j > 0.25$ holds, the three cases for concession policies are

- concessions along the compliance constraint curve (*IIa*),
- concessions by a rightward shift (*IIb*) and
- concessions by an upward shift.

Equations (4.38) and (4.40) hold in general, and the last case can be excluded because upward-shift concessions do not change agent i's marginal utility. The relevant cases are denoted by *IIa* and *IIb*. Note that the effort level of agent i which the best policy specifies for agent j falls short of the Pareto-optimal level:

$$E_i = \left[\delta_i^2 \, \frac{2\alpha_i^2 \alpha_j}{\gamma_i^2 \gamma_j} \right]^{1/3} + \frac{\alpha_i}{\gamma_i} = \left[\delta_i^2 \, \frac{2\alpha_i^2 \gamma_i}{\alpha_j^2 \gamma_j} \right]^{1/3} \frac{\alpha_j}{\gamma_i} + \frac{\alpha_i}{\gamma_i} < \frac{\alpha_i + \alpha_j}{\gamma_i}$$

(4.55)

because

$$\delta_i^2 < 1 \text{ and } \frac{2\alpha_i^2 \gamma_i}{\alpha_j^2 \gamma_j} < 1.$$

Agent *i* is indifferent as to whether options *IIa* or *IIb* are chosen because his/her marginal gains are identical. Agent *j*, however, prefers to make concessions away from his/her optimal policy along the compliance constraint curve.

Let dU_{jIIa} and dU_{jIIb} denote the marginal utility changes which both concession policies produce. According to (3.41), dU_{jIIa} for agent *j*'s optimal policy is given by

$$dU_{jIIa}\left(d_i = \delta_i^2 \right) = \alpha_j \left[1 - \frac{d_i}{\delta_i^2} \right] dE_i = 0 \tag{4.56}$$

which uses the strategic bargaining parameter which Section 3.3 employed. The marginal utility change which option *IIb* implies for marginal moves away from agent *j*'s optimal policy is determined by (4.40). Inserting agent *i*'s effort level for *j*'s optimal policy (see (3.23)) into the lower line of (4.40) gives

$$dU_{jIIb}\left[E_i = \left[\delta_i^2 \, \frac{2\alpha_i^2 \alpha_j}{\gamma_i^2 \gamma_j} \right]^{1/3} + \frac{\alpha_i}{\gamma_i} \right]$$

$$= \left[\alpha_j - \frac{\gamma_i}{\delta_i} \left[\delta_i^2 \, \frac{2\alpha_i^2 \alpha_j}{\gamma_i^2 \gamma_j} \right]^{1/3} \right] dE_i \leq 0 \tag{4.57}$$

because

$$
\frac{\gamma_i}{\delta_i}\left[\delta_i^2\,\frac{2\alpha_i^2\alpha_j}{\gamma_i^2\gamma_j}\right]^{1/3} = \alpha_j\left[\frac{2\alpha_i^2\gamma_i\big/\alpha_j^2\gamma_j}{\delta_i}\right]^{1/3} \ge \alpha_j.
$$

Equation (4.57) demonstrates that the marginal utility change is negative (except for the case $\delta_i = (2\alpha_i^2\gamma_i)/(\alpha_j^2\gamma_j)$) whereas option *IIa* gives a zero change. Equations (4.56) and (4.57) prove that agent j prefers option *IIa* when moving away from his/her optimal policy. Comparing dU_{IIa} and dU_{IIb} determines concession policies in general:

$$
E_j \begin{array}{c}\le\\[-2pt]\overline{}\\[-6pt]>\end{array}\frac{\alpha_i+\alpha_j}{\gamma_j} \Leftrightarrow dU_{jIIa}\begin{array}{c}\ge\\[-6pt]<\end{array}dU_{jIIb} \tag{4.58}
$$

because

$$
\alpha_j - \frac{\gamma_i\gamma_j}{\delta_i\alpha_i}\varepsilon_i\varepsilon_j \begin{array}{c}\ge\\[-6pt]<\end{array}\alpha_j - \frac{\gamma_i}{\delta_i}\varepsilon_i \Leftrightarrow E_j\begin{array}{c}\le\\[-6pt]>\end{array}\frac{\alpha_i+\alpha_j}{\gamma_j}.
$$

In general, agent j favours concessions along the compliance constraint curve until his/her Pareto-optimal level is reached. When this level is reached, concession policies switch to rightward shifts.

The discussion of Case I and Case II has shown that agent j's concession policy starts with concessions along the compliance constraint curves if agent j's optimal policy implies zero transfers, because optimal zero transfers imply efforts which both fall short of the Pareto-optimal ones. Optimal non-zero transfers imply the Pareto-optimal effort level of agent i (if $T_j^* < 0$) or agent j (if $T_j^* > 0$). Thus, if the optimal transfer level is zero, concessions make sense which increase efforts of both agents until one agent's Pareto-optimal level is reached. Then, transfers can play their role in increasing the other agent's effort level until both efforts have reached the Pareto-optimal levels. Table 4.7 summarizes the results of Case I and Case II and completes the discussion of concession policies. Tables 4.6 and 4.7 and the corresponding evaluation for agent i's concession policy cover all relevant cases. However, note again that feasibility aspects have not yet been taken into consideration.

Table 4.7   *Concession policies for* $T_j^* = 0$

	Case I: public good contributor attribute	Case II: seller attribute
	$\dfrac{2\alpha_i^2\gamma_i}{\alpha_j^2\gamma_j} \geq 1, \delta_i \leq \sqrt{\dfrac{\alpha_j^2\gamma_j}{2\alpha_i^2\gamma_i}}$	$\dfrac{2\alpha_i^2\gamma_i}{\alpha_j^2\gamma_j} < 1, \delta_i \leq \dfrac{2\alpha_i^2\gamma_i}{\alpha_j^2\gamma_j}$
	$dE_i > 0, dE_j > 0, dT = 0$	
Concession policies	until agent i's Pareto-optimal level is reached	until agent j's Pareto-optimal level is reached
of agent j	$dE_j > 0, dT < 0, dE_i = 0$	$dE_i > 0, dT > 0, dE_j = 0$
	until both agents' Pareto-optimal levels are reached	

Before turning to feasibility, it should be stressed again that the concession policies pursued by one agent define the concession line: in all relevant cases, the other agent was indifferent as to which relevant option was chosen. Thus, the best concession policy, that is, the policy which maximizes one agent's utility change for a given utility change of the other agent, obviously also defines the concession line.

4.5 FEASIBILITY CONSTRAINTS, STRATEGIC BARGAINING SOLUTIONS AND THE IMPACT OF DISCOUNT FACTOR CHANGES

The last section has developed the concession policies pursued by both agents, which define the concession line. Developing the concession line was based on the assumption that all points under consideration are feasible because both agents' compliance constraints allow realization. This section builds on the development of the concession line and has a threefold intention. First, it will deal with feasibility by considering whether the Pareto-optimal effort levels are feasible under different assumptions. Second, it will emphasize that strategic bargaining plays a similar role as in Chapter 3 because the outcome is determined by the relative impatience of both agents and not by the feasiblity of the Pareto-optimal effort levels. Third, the impact of discount factor changes on the utilities of both agents will be discussed. As the last two points produce results which are quite similar to the results of Chapter 3, they are given less space than in Chapter 3. The discussion of discount factor changes will be restricted to transfer-effort concessions because the impact for other ranges of the concession line is clear or has already been discussed in Chapter 3.

Feasibility Constraints

Until now, only single compliance constraints have been discussed under the assumption that other constraints do not bind. When discussing feasibility, the comprehensive impact of compliance constraints is considered. Hence, this section discusses the conditions under which the compliance constraints of both agents allow certain efforts-transfer combinations to be realized. The discussion of feasibility constraints starts with Case A which gives both agents the public good contributors' attribute. If the Pareto-optimal levels are reachable, the transfer levels on the agents' compliance constraints are given by (4.59):

$$T_j = \frac{\alpha_j^2}{2\gamma_i} - \frac{\delta_i \alpha_i^2}{\gamma_j}, \quad T_i = \frac{\delta_j \alpha_j^2}{\gamma_i} - \frac{\alpha_i^2}{2\gamma_j}. \tag{4.59}$$

The first term of (4.59) is given directly by (4.31), the second term follows from the evaluation for agent i (note that Case A implies different signs for the optimal transfer levels which make both agents receivers). The first term gives the transfer level if concession policies along agent i's compliance constraint have reached Pareto-optimal effort levels of both agents, the second term gives the transfer level along agent j's compliance constraint for the Pareto-optimal levels. Feasibility of the Pareto-optimal level is given if

$$T_i - T_j \geq 0 \tag{4.60}$$

defined by (4.59) holds, which guarantees that the transfers received by one agent along the other agent's compliance constraint do not fall short of the transfers received by this agent along his/her own compliance constraint for the Pareto-optimal levels. Condition (4.60) checks feasibility by comparing transfers enjoyed on one's own compliance constraint with transfers enjoyed on the other agent's compliance constraint. As both agents enjoy the same effort-based utility, feasibility requires that the transfer-based utility on one's own compliance constraint must not exceed the one on the other agent's compliance constraint for given Pareto-optimal effort levels. Otherwise, an agent making concessions from the other agent's compliance constraint to his own compliance constraint would increase his/her bargaining gains which contradicts the behaviour of bargaining gains along the set of efficient agreements. Equations (4.59) and (4.60) define the feasibility constraint

$$\left[\delta_j - \frac{1}{2}\right]\alpha_j^2 \gamma_j + \left[\delta_i - \frac{1}{2}\right]\alpha_i^2 \gamma_i \geq 0. \tag{4.61}$$

Condition (4.61) shows that a necessary condition for feasibility of the Pareto-optimal levels requires that at least one discount factor exceeds 0.5. If both discount factors fall short of 0.5, the Pareto-optimal levels are infeasible and dominated by the compliance constraints. Condition (4.61) also demonstrates the relevance of the benefit and cost parameters for feasibility. The greater these parameters of an agent are, the greater is the impact of his/her discount factor on feasibility. This weighting of discount factors is due to potentially different interests of both agents. For example, an agent with high benefit and cost parameters is relatively more interested in public good provision but his/her own efforts carry high costs. The Pareto-optimal levels benefit him/her overproportionally and increase the transfers which must be paid to the other agent. His/her compliance is more decisive as he/she has to pay more to the other agent (or to receive less from the other agent) and to carry his/her own efforts underproportionally.

According to Table 4.5, Case A comprises four cases, indicated by A_1 to A_4, which represent different discount factor ranges. A comprehensive analysis of all cases can be developed by discussing the limit cases:

$$\delta_i = \sqrt{\frac{\alpha_j^2 \gamma_i}{2\alpha_i^2 \gamma_i}} = \frac{1}{2}\sqrt{\frac{2\alpha_j^2 \gamma_j}{\alpha_i^2 \gamma_i}} > \frac{1}{2},$$

$$\delta_j = \sqrt{\frac{\alpha_i^2 \gamma_i}{2\alpha_j^2 \gamma_j}} = \frac{1}{2}\sqrt{\frac{2\alpha_i^2 \gamma_i}{\alpha_j^2 \gamma_j}} > \frac{1}{2}. \tag{4.62}$$

Note that the limit cases define discount factors which both exceed 0.5. Thus, the limit cases do meet condition (4.61) and prove that realization of the Pareto-optimal levels is possible for all cases A_2 to A_4 and guaranteed for case A_1. A_1 guarantees that both discount factors exceed 0.5, A_2 and A_3 require that at least one discount factor exceeds 0.5. For Case A_4, feasibility of the Pareto-optimal level cannot be excluded because parameter constellations exist which fulfil (4.61). What is known definitely for Case A_4 is that $\delta_i \, \delta_j < 0.5$ because

$$\delta_i \, \delta_j < \sqrt{\frac{\alpha_j^2 \gamma_j}{2\alpha_i^2 \gamma_i}} \sqrt{\frac{\alpha_i^2 \gamma_i}{2\alpha_j^2 \gamma_j}} = \frac{1}{2} \tag{4.63}$$

holds. In general, it depends on the case in hand whether A_2, A_3 or A_4 do not allow the Pareto-optimal levels to be realized.

The condition for feasible points on the contract curve in Chapter 3 required $\delta_i \, \delta_j \geq 0.25$. Comparing (4.61) with this condition is interesting but also somewhat 'unfair' from the viewpoint of the transfer case. In Chapter 3, $\delta_i \, \delta_j \geq 0.25$ guaranteed that one point on the contract curve is feasible. Condition (4.61), however, deals with the feasibility of one specific point because transfers imply a unique effort vector which maximizes total utility. In Chapter 3, $\delta_i \, \delta_j \geq 0.25$ guarantees that one element of a set of Pareto optima is realized. Thus, (4.61) is somewhat more restrictive as it allows only one joint effort policy.

Assume that $\delta_i, \, \delta_j = 0.5$ holds. Setting $\omega := \alpha_i^2 \gamma_i / \alpha_j^2 \gamma_j$ simplifies (4.61) and allows the function Ω to be defined:

$$\left[\frac{1}{4\delta_i} - \frac{1}{2} \right] + \left[\delta_i - \frac{1}{2} \right] \omega \geq 0, \, \Omega(\delta_i, \omega) := \delta_i^2 - \frac{\delta_i}{2}\left[\frac{1}{\omega} + 1 \right] + \frac{1}{4\omega}. \tag{4.64}$$

The first term uses $\delta_i \, \delta_j = 0.25$ and the definition of ω. The second term follows from multiplication of the LHS of the first term by δ_i. A positive (negative) Ω indicates feasibility (infeasibility) of the Pareto-optimal levels. The function Ω has the properties

$$\frac{\partial \Omega}{\partial \omega} = \frac{1}{2\omega^2}\left[\delta_i - \frac{1}{2}\right], \quad \frac{\partial^2 \Omega}{\partial \omega^2} = -\frac{1}{\omega^3}\left[\delta_i - \frac{1}{2}\right],$$

$$\frac{\partial \Omega}{\partial \delta_i} = 2\delta_i - \frac{1}{2}\left[\frac{1}{\omega}+1\right], \quad \frac{\partial^2 \Omega}{\partial \delta_i^2} = 2 > 0, \quad \Rightarrow \delta_i^{\min} = \frac{1}{4}\left[\frac{1}{\omega}+1\right]. \tag{4.65}$$

$\delta_i^{\min}$ denotes the discount factor of agent i which minimizes Ω for any given ω. ω must lie in the range between 0.5 and 2 (see the definition of Case A). Figure 4.5 describes the behaviour of Ω for three ω cases for which $\omega \in \{0.5, 1, 2\}$.

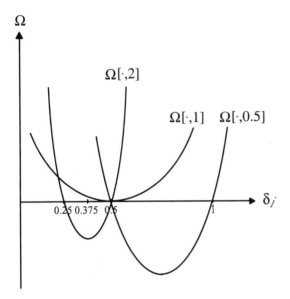

Figure 4.5   *Feasibility of Pareto-optimal levels for bargaining with transfers*

Figure 4.5 shows that feasibility is not guaranteed unless ω is set equal to unity. In all other cases, certain discount factor combinations which give $\delta_i \, \delta_j = 0.25$ fail condition (4.61) because the corresponding parameter constellations emphasize the influence of an impatient agent who prevents the Pareto-optimal levels from being realized. In these cases, one agent is comparably more a seller than a strict contributor to a public good (although still a public good contributor in the sense of the definition). Hence, the Pareto-optimal effort levels specify that he/she realizes few bargaining gains before transfers are paid. This fact makes the non-compliance option more attractive for him/her compared to the case of similar bargaining gains before transfers are paid.

Cases B and C give one agent the seller's attribute. Because Cases B and C are symmetric, this section restricts the discussion to Case B, which gives agent i the seller's attribute. If the Pareto-optimal levels are reachable, the transfer levels on the agents' compliance constraints are

$$T_j = \frac{1}{\delta_i} \frac{\alpha_j^2}{2\gamma_i} - \frac{\alpha_i^2}{\gamma_j}, \quad T_i = \frac{\delta_j \alpha_j^2}{\gamma_i} - \frac{\alpha_i^2}{2\gamma_j}. \tag{4.66}$$

The first term of (4.66) is directly given by (4.42); the second term reformulates (4.31) for agent i. The subscript indicates that transfer levels for the Pareto-optimal levels are considered when agent i or j has made concessions along the other agent's compliance constraint. Feasibility is given if (4.60) holds. Equations (4.60) and (4.66) define the feasibility constraint

$$\frac{1}{2\delta_i} - \delta_j \le \frac{\alpha_i^2 \gamma_i}{2\alpha_j^2 \gamma_j} \Leftrightarrow \delta_j \ge \frac{1}{2\delta_i} - \frac{\alpha_i^2 \gamma_i}{2\alpha_j^2 \gamma_j} \Leftrightarrow \delta_i \ge \frac{1}{2\delta_j + \alpha_i^2 \gamma_i / \alpha_j^2 \gamma_j}. \tag{4.67}$$

The interpretation of (4.67) is not in that way straightforward as it was for (4.61). The intuition behind (4.67) can be made clear by considering the limit cases which distinguish the cases B_1 to B_4 (see Table 4.5):

$$\delta_i = \frac{2\alpha_i^2 \gamma_i}{\alpha_j^2 \gamma_j} = 2\omega, \quad \delta_j = \sqrt{\frac{\alpha_i^2 \gamma_i}{2\alpha_j^2 \gamma_j}} = \frac{1}{2}\sqrt{\frac{2\alpha_i^2 \gamma_i}{\alpha_j^2 \gamma_j}} = \frac{1}{2}\sqrt{2\omega}. \tag{4.68}$$

Equation (4.68) has reformulated both agents' discount factors by the using $\omega = \alpha_i^2 \gamma_i / \alpha_j^2 \gamma_j$. Note that Case B implies $\omega < 1/2$, because

$$\frac{2\alpha_j^2 \gamma_i}{\alpha_j^2 \gamma_j} < 1, \quad \frac{2\alpha_j^2 \gamma_j}{\alpha_i^2 \gamma_i} \ge 1 \Leftrightarrow \frac{2\alpha_i^2 \gamma_i}{\alpha_j^2 \gamma_j} \le 4. \tag{4.69}$$

The second condition is redundant because the first condition which defines the seller's attribute of agent i dominates the second one. Using ω and the limit cases allows the function Φ to be defined:

$$\Phi(\omega) = 2\omega - \frac{1}{\sqrt{2\omega} + \omega}, \quad \frac{d\Phi}{d\omega} = 2 + \frac{(\omega)^{-1/2} + 1}{\left(\sqrt{2\omega} + \omega\right)^2} > 0. \qquad (4.70)$$

A positive (negative) Φ indicates feasibility (infeasibility) of the Pareto-optimal levels. The second line of (4.70) demonstrates that the chances for feasibility are the higher the greater is ω; that is, the less agent i qualifies as a seller. For example,

$$\Phi\left(\omega = \frac{1}{2}\right) = \frac{1}{3}, \Phi\left(\omega = \frac{1}{4}\right) = -0.545$$

shows that the limit case of discount factor combinations may include feasibility of the Pareto-optimal levels if ω lies only slightly below 0.5. Thus, the Pareto-optimal levels are possible for all cases B_1 to B_4 but feasibility is more unlikely the lower ω is.

The lower ω is, that is, the more the problem is changed from a public good problem into a buyer–seller problem, the higher must be agent i's discount factor. The intuition behind (4.70) is now clear: decreasing ω stresses the role of agent i as a seller if the Pareto-optimal levels are to be provided. A seller must receive more transfers the more he/she has to carry efforts compared to the other agent. Carrying comparably more efforts and receiving more transfers makes non-compliance more attractive because the incentive to take the transfer and to refrain from providing this high effort level is stressed. This effect of a decreasing ω can only be compensated for by an increase in agent i's discount factor.

Comparing (4.67) with the feasibility condition (3.30) gives an interesting result although the reservation given for the comparability of both models still applies. Setting $\delta_i \delta_j = 0.25$ for (4.67) requires

$$\delta_i \geq \frac{1}{2\omega} > 1 \qquad (4.71)$$

which can never hold. Thus, $\delta_i \delta_j = 0.25$ excludes feasibility for all B cases which was possible for all A cases. Feasibility of one point on the contract curve of Chapter 3 does not suffice to ensure feasibility of the Pareto-optimal levels if one agent qualifies as a seller.

Feasibility aspects are not only relevant for the Pareto-optimal levels of both agents but also for the Pareto-optimal level of one agent when optimal transfers are zero. If optimal transfers are non-zero, transfers obviously play their role along the whole concession line. For zero optimal transfers, however, Section 4.4 has found that concessions start with effort policies only. In both cases (see Table 4.7), the concession line leaves the transfers on their zero level until one agent's Pareto-optimal level is reached. Hence, transfers can only play a role for these cases if the Pareto-optimal effort level of agent i (j) in Case I (II) is not dominated by the compliance constraints of both agents without transfers. If the other agent's optimal policy implies non-zero transfers, transfers play a role on the concession line. Since transfers may play no role if $T_j^* = T_i^* = 0$ holds, the following discussion is restricted to cases A_4 and B_4.

Case I specifies that both agents increase their effort levels until agent i's Pareto-optimal level is reached. Transfers can only play their role if agent j's compliance constraint curve does not dominate agent i's Pareto-optimal level. The critical δ_j which agent j's discount factor must exceed in order to allow transfers can be determined by computing the intersection of the compliance constraints of both agents at agent i's Pareto-optimal level. If both compliance constraints without transfers intersect at agent i's Pareto-optimal level,

$$\left[d_i \frac{2\alpha_i^2 \alpha_j}{\gamma_i^2 \gamma_j} \right]^{1/3} = \frac{\alpha_j}{\gamma_i} \Leftrightarrow d_i = \frac{\alpha_j^2 \gamma_j}{2\alpha_i^2 \gamma_i} \leq 1 \Rightarrow \delta_j = \frac{1}{4\delta_i^2} \frac{\alpha_j^2 \gamma_j}{2\alpha_i^2 \gamma_i} \geq \frac{1}{4}$$

because

$$\delta_i \leq \sqrt{\frac{\alpha_j^2 \gamma_j}{2\alpha_i^2 \gamma_i}} \Leftrightarrow \frac{1}{4\delta_i^2} \geq \frac{\alpha_i^2 \gamma_i}{4\alpha_j^2 \gamma_j}. \tag{4.72}$$

Condition (4.72) uses (3.36) to compute the bargaining parameter d_i which implies agent i's Pareto-optimal level and (3.63) to compute the intersection in order to determine the minimum δ_j. Condition (4.72), however, contains no contradiction although it assumes that optimal policies of both agents imply zero transfers, that is, that

$$\frac{1}{4} \le \delta_j \le \sqrt{\frac{\alpha_i^2 \gamma_i}{2\alpha_j^2 \gamma_j}} \Rightarrow \frac{1}{4} \le \sqrt{\frac{\alpha_i^2 \gamma_i}{2\alpha_j^2 \gamma_j}} \Leftrightarrow \frac{2\alpha_i^2 \gamma_i}{\alpha_j^2 \gamma_j} \ge \frac{1}{4} \Leftrightarrow \frac{2\alpha_j^2 \gamma_j}{\alpha_i^2 \gamma_i} \le 16$$

$$(4.73)$$

holds. Condition (4.73) demonstrates that there is scope for a δ_j which starts concession policies without transfers and changes to a transfer-based concession policy because the restrictions imposed by (4.72) do not violate the assumptions made for Case A_4.

Case II specifies that both agents increase their effort levels until agent j's Pareto-optimal level is reached. The critical δ_j can be determined along a similar line of reasoning for agent j's Pareto-optimal effort level:

$$\frac{1}{\delta_i} \frac{\gamma_i}{2\alpha_i} \left[d_i \frac{2\alpha_i^2 \alpha_j}{\gamma_i^2 \gamma_j} \right]^{2/3} = \frac{\alpha_i}{\gamma_j} \Leftrightarrow d_i = \frac{2\alpha_i^2 \gamma_i}{\alpha_j^2 \gamma_j} \delta_i^3$$

$$\Rightarrow \delta_j = \frac{1}{4} \sqrt{\frac{1}{\delta_i} \frac{2\alpha_i^2 \gamma_i}{\alpha_j^2 \gamma_j}} < \frac{1}{4} \sqrt{\frac{1}{\delta_i}} \qquad (4.74)$$

Equation (4.74) does not contradict the assumption of zero optimal transfers for both agents:

$$\frac{1}{4} \sqrt{\frac{1}{\delta_i} \frac{2\alpha_i^2 \gamma_i}{\alpha_j^2 \gamma_j}} \le \delta_j \le \sqrt{\frac{\alpha_i^2 \gamma_i}{2\alpha_j^2 \gamma_j}} \Rightarrow \delta_i \ge \frac{1}{2} \Rightarrow \frac{1}{2} \le \delta_i \le \frac{2\alpha_i^2 \gamma_i}{2\alpha_j^2 \gamma_j} < 1.$$

$$(4.75)$$

The first equation of (4.75) gives the range of agent j's discount factors which let concession policies start without transfers and switch to transfers as agent j's Pareto-optimal level is reached. The second equation gives a necessary condition for the first equation and it shows that this condition does not conflict with the restriction imposed on agent i's discount factor. Thus, (4.74) is compatible with the conditions which defined Case B_4.

Condition (4.75) completes the discussion of feasibility. Feasibility, however, lets a specific agreement on the concession line be a candidate only for the strategic bargaining solution. The strategic bargaining solution is determined by equalizing the elasticity term ρ and $-\ln\delta_i/\ln\delta_j$. As Table 4.5 indicates twelve possible cases and the feasibility discussion has demonstrated the different possibilities, the behaviour of ρ is hard to characterize in general. But one may discuss the behaviour of ρ along the two compliance constraints and the purely transfer-based concessions which have been developed in this chapter. Additionally, it should be clear that the feasibility conditions have a direct implication for the behaviour of ρ at the points under consideration. All concession policy changes which are summarized in Tables 4.6 and 4.7 were determined by comparing dU_i and dU_j. A change in concession policies coincides with identical dU_i and dU_j at the point of change from an old to a new concession policy. When dU_i and dU_j are equal for both concession policies and the effort vector is fixed, that is, the effort-based utility is given, ρ is obviously identical for both concession policies at this point. Therefore, if the respective feasibility conditions hold, the elasticity term ρ does not jump along the concession line. As long as ρ does not jump, a specific $-\ln\delta_i/\ln\delta_j$ will be shown to correspond to a unique ρ and a specific ρ will be shown to correspond to a unique $-\ln\delta_i/\ln\delta_j$. If a feasibility constraint binds, ρ will jump from one to the other concession policy. A jump implies that a specific $-\ln\delta_i/\ln\delta_j$ corresponds to a unique ρ but a specific agreement can be implied by different combinations of $-\ln\delta_i/\ln\delta_j$.

Strategic Bargaining Solutions and the Impact of Discount Factor Changes

The discussion of the behaviour of ρ will be restricted to the upward-shift concessions of Case A, the rightward-shift concessions of Case B and the purely transfer-based policy. Other concession policies leave transfers unchanged and were discussed explicitly in Chapter 3. Along the upward-shift concession line of Case A, using (4.27) and (4.36) gives

$$\rho_{US} = \frac{1-\delta_i}{\delta_i - D_{in}} \frac{\frac{\alpha_j^2}{2\gamma_i} + (2\delta_i - D_{in})\frac{D_{in}\alpha_i^2}{2\gamma_j} + V_j'}{(1-\delta_i)\frac{D_{in}\alpha_i^2}{\gamma_j} + V_i'},$$

$$\lim_{E_j \to \frac{\delta_i\alpha_i + \alpha_j}{\gamma_j}} \rho_{US} = \lim_{D_{in} \to \delta_i} \rho_{US} = -\infty,$$

$$\rho_{US}\left(E_i = \frac{\alpha_i + \alpha_j}{\gamma_i}, E_j = \frac{\alpha_i + \alpha_j}{\gamma_j}\right) = \rho_{US}(D_{in} = 1) = -\frac{V_j}{V_i} = \rho_T. \qquad (4.76)$$

The subscript *US* denotes the upward shift, the subscript *T* denotes the exclusively transfer-based concessions. V_i' (V_j') is the non-cooperative utility level of agent i (j), and D_{in} denotes the bargaining parameter which (4.35) has introduced. Equation (4.76) gives the behaviour of ρ if agent j's optimal policy implies non-zero transfers. For optimal zero transfers, the results of Chapter 3 carry directly over on the part of the concession line which does not change the zero transfer level. Equation (4.76) shows that the optimal policies induce an infinitely negative ρ which can only be met by $\delta_j = 1$. The third line demonstrates that ρ does not jump when the upward-shift policy is replaced by pure transfer policies because ρ_{US} is equal to ρ_T at the Pareto-optimal effort levels. Additionally, one may derive from (4.36) that

$$\frac{dU_i}{dD_{in}} = (1-\delta_i)\frac{\alpha_i^2}{\gamma_j} > 0, \quad \frac{dU_j}{dD_{in}} = (\delta_i - D_{in})\frac{\alpha_i^2}{\gamma_j} < 0 \Rightarrow \frac{dU_i}{dU_j} = \frac{1-\delta_i}{\delta_i - D_{in}} < 0,$$

$$\frac{d(dU_i/dU_j)}{dD_{in}} = \frac{1-\delta_i}{(\delta_i - D_{in})^2} > 0,$$

$$\frac{d(V_j/V_i)}{dD_{in}} = \frac{(dU_j/dD_{in})V_i - (dU_i/dD_{in})V_j}{V_i^2} < 0,$$

$$\frac{d\rho_{US}}{dD_{in}} = \frac{d(dU_i/dU_j)}{dD_{in}}\frac{V_j}{V_i} + \frac{d(V_j/V_i)}{dD_{in}}\frac{dU_i}{dU_j} > 0. \tag{4.77}$$

Equation (4.77) shows that ρ increases along the upward-shift concessions. Hence, any $-\ln\delta_i/\ln\delta_j$ which implies a solution on this part of the concession line implies one and only one solution.

Along the rightward-shift concessions of Case B, using (4.40) and (4.46) gives

$$\rho_{RS} = \frac{1-\delta_i}{\delta_i - D_{ip}} D_{ip} \frac{\alpha_i^2/2\gamma_j + \left[1-(D_{ip}/2\delta_i)\right]D_{ip}\alpha_j^2/\gamma_i + V_j'}{D_{ip}^2\alpha_j^2(1-\delta_i)/2\gamma_i\delta_i + V_i'}$$

$$\lim_{E_i \to \frac{\delta_i\alpha_i + \alpha_j}{\gamma_i}} \rho_{RS} = \lim_{D_{ip} \to \delta_i} \rho_{RS} = -\infty$$

$$\rho_{RS}\left(E_i = \frac{\alpha_i + \alpha_j}{\gamma_i}, E_j = \frac{\alpha_i + \alpha_j}{\gamma_j}\right) = \rho_{RS}\left[D_{ip} = 1\right] = -\frac{V_j}{V_i} = \rho_T. \tag{4.78}$$

The subscript *RS* denotes the rightward shift. Equation (4.78) gives the behaviour of ρ for non-zero optimal policies for agent *i*. For optimal zero transfers, the concession line of Chapter 3 replaces the first part. Equation (4.78) demonstrates that the optimal policies induce an infinitely negative ρ which can only be met by $\delta_j = 1$ as a strategic bargaining solution. The third line proves that ρ does not jump when the rightward-shift policy is replaced by transfer policies. Additionally, one may derive from (4.46) that

$$\frac{dU_i}{dD_{ip}} = \frac{\alpha_j^2}{\gamma_i}\frac{1-\delta_i}{\delta_i} > 0, \quad \frac{dU_j}{dD_{ip}} = \left[1 - \frac{D_{ip}}{\delta_i}\right]\frac{\alpha_j^2}{\gamma_i} < 0 \Rightarrow \frac{dU_i}{dU_j} = \frac{1-\delta_i}{\delta_i - D_{ip}} < 0,$$

$$\frac{d\left(dU_i/dU_j\right)}{dD_{ip}} = \frac{1-\delta_i}{\left(\delta_i - D_{ip}\right)^2} > 0,$$

$$\frac{d\left(V_j/V_i\right)}{dD_{ip}} = \frac{\left(dU_j/dD_{ip}\right)V_i - \left(dU_i/dD_{ip}\right)V_j}{V_i^2} < 0,$$

$$\frac{d\rho_{RS}}{dD_{ip}} = \frac{d\left(dU_i/dU_j\right)}{dD_{ip}}\frac{V_j}{V_i} + \frac{d\left(V_j/V_i\right)}{dD_{ip}}\frac{dU_i}{dU_j} > 0. \tag{4.79}$$

Equation (4.79) shows that ρ increases along the rightward-shift concessions. Hence, any $-\ln\delta_i/\ln\delta_j$ which implies a solution on this part of the concession line implies one and only one solution.

As transfers enter the utilities of both agents by simple addition to the effort-based utility, describing bargaining gains and marginal bargaining gains is straightforward for the exclusively transfer-based concession policy:

$$dU_i = -dU_j = 1,$$

$$U_i = \frac{\alpha_i^2}{\gamma_j} - \frac{\alpha_j^2}{2\gamma_i} + \tau, \quad U_j = \frac{\alpha_j^2}{\gamma_i} - \frac{\alpha_i^2}{2\gamma_j} - \tau, \quad \rho_T = -\frac{V_j}{V_i}. \tag{4.80}$$

τ denotes the bargaining parameter for transfers and lies in the limits (4.31) has defined for Case A and (4.42) has defined for Cas B. From (4.80) it is clear that

$$\frac{d\rho_T}{d\tau} = \frac{V_i + V_j}{V_i^2} > 0. \tag{4.81}$$

Equation (4.81) shows that ρ increases along the exclusively transfer-based concession policy, and any $-\ln\delta_i/\ln\delta_j$ which implies a solution on this part of the concession line implies one and only one solution. Since (4.76) and (4.78) have shown that ρ is equal for two concession policies when one policy switches to another at the Pareto-optimal effort levels, uniqueness is guaranteed because ρ increases along the whole concession policy (note that the discussion of D_{in} and D_{ip} is sufficient because D_{jp} and D_{jn} give a similar picture). If the Pareto-optimal effort levels are not feasible, ρ jumps at this point but uniqueness is still guaranteed because ρ increases over the whole range.

Equations (4.76), (4.78), (4.80) and the descriptions given in Chapter 3 can determine every concession line which every case in Table 4.5 implies. Since uniqueness was also guaranteed for the compliance constraints of Chapter 3 and ρ does not jump when the concession policy switches (if feasible), uniqueness is guaranteed for all conceivable cases. In all cases, the absolute value of the elasticity which is given by these descriptions must be equal to $\ln\delta_i/\ln\delta_j$ in order to determine the strategic bargaining solution.

As in Section 3.3, the strategic bargaining solution can also be given as

$$\ln\delta_j \frac{dV_i(\cdot)}{V_i(\cdot)} + \ln\delta_i \frac{dV_j(\cdot)}{V_j(\cdot)} = 0.$$

Consider first strategic bargaining for which only D_{in} describes the relevant range. From (4.36),

$$\frac{\partial U_i}{\partial \delta_i}(D_{in}) < 0, \frac{\partial U_i}{\partial D_{in}}(D_{in}) > 0, \frac{\partial U_j}{\partial \delta_i}(D_{in}) > 0, \frac{\partial U_j}{\partial D_{in}}(D_{in}) < 0 \tag{4.82}$$

can be derived. For discussing the impact of discount factor changes for the range D_{in} describes, the bargaining gains will be dealt with as functions of D_{in} and δ_i. The best policy for agent j is $D_{in} = \delta_i$ which determines bargaining gains according to (4.14). Differentiating (4.14) with respect to δ_i gives

$$\frac{dU_i}{d\delta_i}(\delta_i) < 0, \frac{dU_j}{d\delta_i}(\delta_i) > 0. \tag{4.83}$$

Equation (4.83) shows that an increase of the discount factor which defines the compliance constraint increases the utility of his/her opponent and decreases his/her own utility at this point. The same result was demonstrated in Section 3.4.

Integration for the range D_{in} describes gives

$$F_{US}(\delta_i, D_{in}) :=$$
$$\ln\delta_j\{\ln V_i(D_{in}) - \ln V_i(\delta_i)\} + \ln\delta_i\{\ln V_j(D_{in}) - \ln V_j(\delta_i)\} = 0. \quad (4.84)$$

Differentiating F_{US} gives

$$\frac{\partial F_{US}}{\partial \delta_i} = \ln\delta_j\left\{\frac{\partial V_i(D_{in})/\partial\delta_i}{V_i(D_{in})} - \frac{dV_i(\delta_i)/d\delta_i}{V_i(\delta_i)}\right\}$$

$$+ \ln\delta_i\left\{\frac{\partial V_j(D_{in})/\partial\delta_i}{V_j(D_{in})} - \frac{dV_j(\delta_i)/d\delta_i}{V_j(\delta_i)}\right\}$$

$$+ \frac{\ln V_j(D_{in}) - \ln V_j(\delta_i)}{\delta_i},$$

$$\frac{\partial F_{US}}{\partial D_{in}} = \ln\delta_j\frac{\partial V_i(D_{in})/\partial D_{in}}{V_i(D_{in})} + \ln\delta_i\frac{\partial V_j(D_{in})/\partial D_{in}}{V_j(D_{in})}, \quad (4.85)$$

both ambiguous in sign, and

$$\Rightarrow \frac{dD_{in}}{d\delta_i}, \frac{dV_i(D_{in})}{d\delta_i} = \frac{\partial V_i(D_{in})}{\partial\delta_i} + \frac{\partial V_i(D_{in})}{\partial D_{in}}\frac{dD_{in}}{d\delta_i},$$

$$\frac{dV_j(D_{in})}{d\delta_i} = \frac{\partial V_j(D_{in})}{\partial\delta_i} + \frac{\partial V_j(D_{in})}{\partial D_{in}}\frac{dD_{in}}{d\delta_i},$$

all ambiguous in sign.

Equation (4.85) shows that the effect of changing agent i's discount factor is ambiguous in this range (and – due to the logarithmic terms – an explicit computation does not resolve ambiguity).

The same results hold for the range which D_{ip} describes. Let the bargaining gains now be dealt with as functions of D_{ip} and δ_i:

$$\frac{\partial U_i}{\partial \delta_i}\left(D_{ip}\right) < 0, \frac{\partial U_i}{\partial D_{ip}}\left(D_{ip}\right) > 0, \frac{\partial U_j}{\partial \delta_i}\left(D_{ip}\right) > 0, \frac{\partial U_j}{\partial D_{ip}}\left(D_{ip}\right) < 0 \qquad (4.86)$$

For the best policy for agent j, (4.87) holds:

$$\frac{dU_i}{d\delta_i}\left(\delta_i\right) < 0, \frac{dU_j}{d\delta_i}\left(\delta_i\right) > 0. \qquad (4.87)$$

Integration for the range D_{ip} describes gives

$$F_{RS}\left(\delta_i, D_{ip}\right) := $$
$$\ln \delta_j \left\{ \ln V_i\left(D_{ip}\right) - \ln V_i\left(\delta_i\right) \right\} + \ln \delta_i \left\{ \ln V_j\left(D_{ip}\right) - \ln V_j\left(\delta_i\right) \right\} = 0. \qquad (4.88)$$

Differentiating F_{RS} gives

$$\frac{\partial F_{RS}}{\partial \delta_i} = \ln \delta_j \left\{ \frac{\partial V_i\left(D_{ip}\right)/\partial \delta_i}{V_i\left(D_{ip}\right)} - \frac{dV_i\left(\delta_i\right)/d\delta_i}{V_i\left(\delta_i\right)} \right\}$$

$$+ \ln \delta_i \left\{ \frac{\partial V_j\left(D_{ip}\right)/\partial \delta_i}{V_j\left(D_{ip}\right)} - \frac{dV_j\left(\delta_i\right)/d\delta_i}{V_j\left(\delta_i\right)} \right\}$$

$$+ \frac{\ln V_j\left(D_{ip}\right) - \ln V_j\left(\delta_i\right)}{\delta_i},$$

$$\frac{\partial F_{RS}}{\partial D_{ip}} = \ln \delta_j \frac{\partial V_i\left(D_{ip}\right)/\partial D_{ip}}{V_i\left(D_{ip}\right)} + \ln \delta_i \frac{\partial V_j\left(D_{ip}\right)/\partial D_{ip}}{V_j\left(D_{ip}\right)}, \qquad (4.89)$$

both ambiguous in sign, and

$$\Rightarrow \frac{dD_{ip}}{d\delta_i}, \frac{dV_i\left(D_{ip}\right)}{d\delta_i} = \frac{\partial V_i\left(D_{ip}\right)}{\partial \delta_i} + \frac{\partial V_i\left(D_{ip}\right)}{\partial D_{ip}} \frac{dD_{ip}}{d\delta_i},$$

$$\frac{dV_j\left(D_{ip}\right)}{d\delta_i} = \frac{\partial V_j\left(D_{ip}\right)}{\partial \delta_i} + \frac{\partial V_j\left(D_{ip}\right)}{\partial D_{in}} \frac{dD_{ip}}{d\delta_i},$$

all ambiguous in sign.

Thus, the discussion of discount factor changes has demonstrated that the ambiguity results of Chapter 3 carry over to Chapter 4. If the compliance constraint of an agent is binding, the concession line is defined by the upward-shift concessions, the rightward-shift concessions, concessions along the compliance constraint curve or a combination of one of the first two concession policies with concessions along the compliance constraint curve. All four options cover all cases of a binding compliance constraint in Table 4.5 because the first two options cover all non-zero optimal transfer cases and the last two options cover all zero optimal transfer cases.

Chapter 3 has demonstrated that the impact of a discount factor variation on the agents' utilities is ambiguous if this discount factor defines the relevant part of the concession line. Therefore, Chapter 3, (4.85) and (4.89) have shown that this ambiguity result also holds for the transfer case as all cases are covered which alter the discount factor defining the relevant compliance constraint. Obviously, the effect of a discount factor change is unambiguous if this discount factor does not change the part of the concession line under consideration. In this case, the bargaining effect increases (decreases) the bargaining gains of the agent whose relative patience is increased (decreased).

Bargaining with Transfers vs. Bargaining without Transfers

The previous sections and this section have demonstrated the increase in both variety and complexity which the introduction of transfers produces for a simple model of strategic bargaining for self-enforcing contracts. In general, it is not possible to determine whether strategic bargaining with transfers improves on the outcome without transfers because utilities and marginal utilities are different along the compliance constraints and for the Pareto-optimal levels. If the solution specifies the Pareto-optimal levels, bargaining with transfers may be supposed to improve on the outcome without transfers because the maximum total bargaining gains can be redistributed. But redistribution of a higher total utility does not necessarily imply higher utilities for both agents, as the strategic bargaining solution for the Pareto-optimal levels implies different total and marginal utilities from those on the contract curve.

In order to shed some light on the differences between bargaining with and without transfers, one may consider two examples. The first case is given by two agents with rather similar benefit and cost parameters. In this case, both agents are public good contributors, and the payoff frontiers for bargaining with transfers and bargaining without transfers are given by Figure 4.6. Figure 4.6 assumes that the Pareto-optimal effort levels are feasible. Point 1 depicts the strategic bargaining solution without transfers. Since costs and benefits are similar, the payoff frontier without transfers is tangential to the payoff frontier with transfers at the point for which $T = 0$. Because $dU_i/dU_j = -1$ for exclusively transfer-based concessions, the slope of the payoff frontier with transfers is linear in this range.

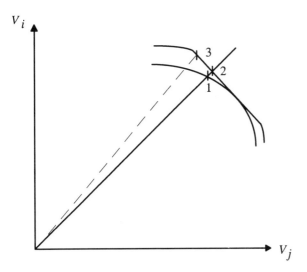

Figure 4.6   Bargaining without and with transfers I

The slope of the unbroken line through 1 gives V_i/V_j. This ratio holds for point 2 as well, but $-dU_i/dU_j$ is greater at 2 (with transfers) than at 1 (without transfers). Hence, 2 cannot give the strategic bargaining solution with transfers if 1 gives the strategic bargaining solution without transfers. Since $-dU_i/dU_j$ is constant in this range (with transfers)

$$-\rho = \frac{-dU_i/dU_j}{V_i/V_j}$$

can be adjusted only by changes in V_i/V_j. The corresponding bargaining solution with transfers must therefore increase V_i/V_j in order to equalize ρ and $-\ln\delta_i/\ln\delta_j$. One might find the solution at point 3 for which the slope of the broken line gives V_i/V_j. But point 3 gives agent j lower bargaining gains compared to point 1. Although the payoff frontier with transfers Pareto dominates the payoff frontier without transfers, one of the agents might lose by allowing transfers.

If costs and benefits are substantially different, the chances of improvement for both agents are increased. In this case, the Pareto-optimal effort levels do not belong to the set of efficient agreements without transfers (see (3.4) and (4.22)). Hence, the payoff frontier without transfers is far away from the payoff frontier with transfers. Figure 4.7 depicts this case.

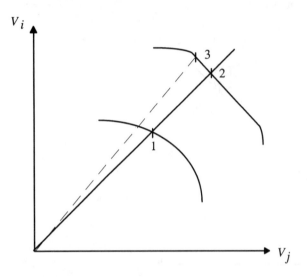

Figure 4.7   Bargaining without and with transfers II

Figure 4.7 gives a similar payoff frontier without transfers as in Figure 4.6. At point 1, the strategic bargaining solution without transfers, ρ is equal to ρ at point 1 in Figure 4.6, and the slope of the line through 3 is the same as through 3 in Figure 4.6. Since the payoff frontiers with transfers specifies significant possible improvements, point 3 benefits both agents. Thereby, Figures 4.6 and 4.7 demonstrate that substantial differences in the cost and benefit parameters may lead to an increase in the bargaining gains for both agents. Hence, it seems that transfers are not capable of reducing compliance costs for both agents but that they are able to exploit efficiency gains. These efficiency gains are larger, the lower are one agent's cost and benefit parameters compared to the other agent.

4.6   APPLICATION II: INTERNATIONAL ENVIRONMENTAL POLICY

One classical application of strategic bargaining for self-enforcing contracts with transfers is international environmental policies, whose relevance has become more pronounced because of the risks of global change. Problems like the accumulation of greenhouse gases in the atmosphere, the loss of biological diversity and the shrinking ozone shelter have shifted the focus of public concern from local hot spots to international and global pollution problems. Expanding the pollution-concerned area did not merely create a new spatial problem. The new economic problem associated with the use of an international environmental resource is due to the lack of a central authority which can enforce a specific environmental resource use. From an international perspective, regulation must be grounded on the voluntary participation of those countries which use and exploit the environmental resource. The sovereignty of countries requires that international environmental agreements must guarantee that every national participant is at least not worse off by fulfilling the agreement than by breaking it.

Because restricting international pollution is a public good, environmentalists fear that the strategy of free riding will become the dominant one of all countries and will make environmental quality significantly worse or will even endanger human living conditions. Economists discussing international pollution problems are not pessimistic about that, although they do not claim that the cooperative outcome will always prevail. Most of the respective literature concentrates on the possible formation of stable coalitions which are a subset of resource-using countries (see Barrett, 1991, 1992; Bauer, 1992; Carraro and Siniscalco, 1993; Heal, 1992). Members of a coalition are supposed to introduce environmental policies by maximizing the welfare of the coalition. Stability requires that every sovereign member country is always better off by remaining in the coalition than by leaving it and every outsider is always worse off by joining the coalition. Partial cooperation can emerge if a stable coalition contains more than one member.

These papers deal with the notion of sovereignty only in terms of voluntary participation. When countries have decided to sign an environmental treaty, they are supposed to behave compliantly and to be able to sustain credibly the agreed-upon reduction efforts. This section addresses the problem that countries need not behave compliantly because they can always repudiate the demands of other countries as an interference in their sovereign affairs. Accordingly, this section assumes that the agreed-upon reduction plans must be self-enforcing and that all countries decide simultaneously whether they will fulfil the agreed-upon policy.

This section will develop the strategic bargaining solutions for several self-enforcing contracts on an empirical basis. It takes the 'European perspective' and considers the result of strategic bargaining for a self-enforcing emission reduction contract between the European Union (EU) and different partners. Joint reduction policies concern the reduction of carbon dioxide (CO_2) emissions which contribute significantly, though not exclusively, to the greenhouse effect. Since the EU has revealed a strong preference for joint environmental policies and has launched many international policy initiatives, this section assumes that the EU's institutional toughness is relatively high. Consequently, the discount factor of the EU is assumed to be equal to 0.95 when annual periods are considered.

The empirical basis of the strategic bargaining results are benefit and cost functions as given in Botteon and Carraro (1995). This paper estimates the damage of CO_2 emissions using as a benchmark the increased mortality rates in the respective regions. Table 4.8 summarizes the relevant parameters for developing the strategic bargaining solutions. All estimates are on an annual basis, and thus all discount factor specifications refer to an annual basis as well. For compliance revelation, it is assumed that non-compliance of a country will be revealed within one year. Thus, any punishment of a non-compliant country will start one year later.

Table 4.8 shows that the estimate assumes linear damages which depend on the world emissions and quadratic abatement costs which depend on the country emissions. This assumption fits the theoretical model of this chapter. As avoided damage equals gross benefits, the bargaining gains for any country k out of l ($l = 5$ in the case under consideration) countries are given by

$$U_k = \alpha_k \sum_l \left(R_l - R_l' \right) - \frac{\gamma_k}{2} \left(R_k - R_k' \right)^2 \tag{4.90}$$

where R denotes realized reductions and the prime denotes the non-cooperative reduction level. As the numbers in Table 4.8 induce non-cooperative reduction levels which are negligibly small, in particular when compared to the emission level, the non-cooperative levels may be approximated by zero, and instead of (4.90),

$$U_k = \alpha_k \sum_l R_l - \frac{\gamma_k}{2} R_k^2 \tag{4.91}$$

will define the bargaining gains.

Table 4.8 Country data set

Regions	European Union (EU)	Japan	USA and Canada (NA)	Eastern Europe and Russia (EER)	China and India
CO_2 emissions in million tons	815	238	1320	1263	722
Domestic damage in US dollars per ton of CO_2 emissions $[\alpha_k]$	26.01	8.51	18.83	25.1	129.84
Slope of the marginal abatement cost function per ton of country emissions $[\gamma_k]$	1.02	4.89	0.54	0.44	0.62

Source: Botteon and Carraro (1995, p. 19).

Employing the numbers given in Table 4.10 should not induce uncritical appraisal of this estimate (for an overview of alternative estimates, see Fankhauser, 1995). In particular, the high damage for China and India is surprising, but is due to estimating damage levels using as a benchmark an increase in mortality. Countries like China and India are heavily populated and an increase in mortality would induce high damage levels. Because of this result, this section will refrain from considering strategic bargaining with the region of China and India. Based on the data of Table 4.8, bargaining would give this region the public good contributor attribute and the EU the seller attribute. It is barely conceivable that strategic bargaining between these two regions would imply that China and India should pay transfers to the EU in every case. Thus, realistic bargaining results which use this data set are restricted to contracts of the EU with Japan, the USA and Canada (NA) or Eastern Europe and Russia (EER).

Determining the attributes of all three potential partners shows that Japan is a public good contributor whereas both NA and EER are sellers. In addition to the data of Table 4.8, it is assumed that Japan's and NA's discount factors belong to the set $\{0.8, 0.85, 0.9, 0.95\}$ and EER's discount factor belongs to the set $\{0.2, 0.3, 0.4, 0.5, 0.6, 0.7, 0.8\}$. This assumption specifies that the degree of institutional toughness is significantly higher in Japan and in NA compared to EER. This is quite a reasonable assumption when the current institutional weakness during the adjustment process of the EER region is taken into account.

According to the functional specification, there is no leakage when two partners join a self-enforcing contract because other regions do not decrease their reductions as a reaction to the agreed-upon reductions of those two partners. The following Tables 4.9–4.11 summarize the strategic bargaining solutions for all cases. The second column of these tables gives the applied bargaining parameter and $R_k{}^*$ denotes the Pareto-optimal reduction level of country k if cooperation is restricted to those two countries under consideration.

The tables indicate that the European Union never receives transfers. This result is due to the relatively high damage parameter of the EU which implies high bargaining gains. Strategic bargaining for a self-enforcing contract with Japan and NA employs the Pareto-optimal levels except in one case for NA where NA's reduction level falls short of the Pareto-optimal one.

Table 4.9 Cooperation between EU (i) and Japan (j) (Japan is a public good contributor)

δ_j	d_i, D_i or τ	R_i	R_j	$R_i^* - R_i$	$R_j^* - R_j$	Transfers from i to j	U_i	U_j
0.8	τ	33.84	7.06	0	0	17.74	85.11	19.56
0.85	τ	33.84	7.06	0	0	23.28	79.56	25.11
0.9	τ	33.84	7.06	0	0	32.45	70.40	34.27
0.95	τ	33.84	7.06	0	0	50.51	52.34	52.34

Note: All reductions are million tons, all transfer levels and all utilities are million US dollars, d_j, D_j, τ: relevant bargaining parameters.

Table 4.10 Cooperation between EU (i) and NA (j) (NA is a seller)

δ_j	d_i, D_i or τ	R_i	R_j	R_i^* $-R_i$	R_j^* $-R_j$	Transfers from i to j	U_i	U_j
0.8	D_j	43.961	82.621	0	0.416	1660.55	646.258	153.911
0.85	τ	43.961	82.621	0	0	470.762	608.245	191.971
0.9	τ	43.961	82.621	0	0	540.806	538.200	261.015
0.95	τ	43.961	82.621	0	0	678.899	400.108	400.108

Table 4.11 Cooperation between EU (i) and EER (j) (EER is a seller)

δ_j	d_i, D_i or τ	R_i	R_j	R_i^* $-R_i$	R_j^* $-R_j$	Trans-fers from i to j	U_i	U_j
0.2	d_j	27.425	63.673	22.683	52.486	0	366.451	320.512
0.3	d_j	29.359	68.538	20.749	47.621	0	480.012	324.197
0.4	d_j	31.882	74.112	18.225	42.047	0	581.153	309.506
0.5	d_j	35.048	80.384	15.060	35.775	0	673.692	282.163
0.6	d_j	39.009	87.455	11.099	28.704	0	759.335	245.101
0.7	D_j	50.108	106.144	0	10.015	1955.36	828.244	227.290
0.8	τ	50.108	116.159	0	0	352.524	876.194	201.408

Note: All reductions are million tons, all transfer levels and all utilities are million
US dollars, d_j, D_j, τ: relevant bargaining parameters.

Cooperation with EER shows that transfers are paid only in two cases. The relatively low discount factors imply concessions along the compliance constraint curve of EER. In this range, the bargaining gains of EER are at a maximum because the compliance effect obviously dominates the bargaining effect in the range between 0.3 and 0.6. Transfers only play a role for the two highest discount factor assumptions.

It is interesting to compare cooperation with NA with cooperation with EER. In the political arena, partial cooperation is seen as a first step for worldwide cooperation. If worldwide cooperation were to start with cooperation between two regions, comparing both scenarios prepares the ground for solving the problem of team selection. For $\delta_j = 0.8$ in NA and EER, the EU is better off by joining a self-enforcing reduction contract with EER. Even the total bargaining gains of this cooperation surmount the total bargaining gains of cooperation between EU and NA. This result is due to the data which specify significantly lower marginal abatement costs for EER than for NA.

But lower discount factors of EER strengthen the compliance effect of the strategic bargaining solution. If the discount factor of EER is 0.2, the EU is always better off by joining a self-enforcing reduction contract with NA. For all other cases, the preference depends on the relationship of both regions' discount factors: high discount factors of NA make preference of the EU for NA unlikely because only the bargaining effect applies, which decreases EU's bargaining gains for increased discount factors of NA. Conversely, low discount factors of EER make preference of the EU for EER unlikely because the compliance effect dominates, which decreases EU's bargaining gains for decreased discount factors of EER. Thus, this example demonstrates that the preference for a partner depends crucially on the strength of both the bargaining effect and the compliance effect.

The discussion of these examples has shown that transfers may play a role in international environmental policies only if the partners are not too different with respect to their institutional performance. This result also holds for a partner which reveals the seller attribute. Table 4.11 demonstrates that the role of transfers is severely restricted although the beneficial impact of transfers is potentially high between these two partners. For low discount factors of the country which has the seller attribute, however, refraining from transfers is the outcome of a bargaining process in which each country maximizes its individual bargaining gains subject to potential delay costs and compliance of the other partner. Thus, one may not expect the introduction of transfers when joint environmental policies of such two partners are to be introduced.

4.7 APPLICATION III: TRADE POLICY

The discussion of trade issues has a long tradition in economics. At least on theoretical grounds, free trade has been strongly advocated as it was shown that free trade maximizes the world welfare. This section will discuss trade policy from the standpoint that every trade agreement between two countries must be self-enforcing. After summarizing the causes for the profitability of free trade, it demonstrates why the unilateral imposition of trade barriers may improve the welfare of a single country. This improvement makes international trade policies likely to face a prisoners' dilemma situation in a non-cooperative environment. Then, this section will turn to strategic bargaining for a self-enforcing contract by considering the chances that such a contract will abolish all trade barriers.

The standard models of international trade assume that domestic production factors are intersectorally mobile but internationally immobile. Production in all sectors is perfectly competitive and employs linear-homogeneous production technologies so that the value of production equals factor income. This section adopts the usual assumptions that the world consists of two countries i and j each of which consumes and produces two goods 1 and 2 which are produced in the respective two sectors. Additionally, it is assumed that no country specializes completely in the production of one good under all scenarios which are considered.

Demand is explained by means of a homothetic utility function V:

$$\forall k \in \{i,j\}: \quad V_k = V_k\left[c_k^1, c_k^1\right]. \tag{4.92}$$

c denotes consumption in the country (which is denoted by the subscript) of the good which is denoted by the superscript. Production possibilities in country k are based on linear-homogenous production functions in each sector. Production is subject to a resource endowment constraint so that any resource input for producing good 1 implies opportunity costs of not being able to produce good 2. The production functions $q_k^1(.)$, $q_k^2(.)$ and the resource constraints imply a transformation function which specifies the points of efficient production. This transformation function maximizes the production in one sector for a given (feasible) production of the other sector. Let the set of production levels which define the transformation function be denoted by T_k for country k. T_k is defined by

$$\forall k \in \{i,j\}: \quad T_k = T_k\left[q_k^1, q_k^1\right] = 0, \quad \frac{\partial T_k}{\partial q_k^1}, \frac{\partial T_k}{\partial q_k^2} > 0. \tag{4.93}$$

Linear-homogeneous production functions are well known to fulfil the additional condition that the frontier of the transformation function is concave and continuously differentiable in the q^1–q^2 space if the elasticity of substitution is not zero. This assumption and (4.93) are necessary (though not sufficient) assumptions to induce interior solutions.

This section will use good 2 as a numeraire. The assumptions of perfectly competitive market structures, homothetic utility functions, concave transformation functions and no externalities are completely consistent with the assumptions made by Arrow and Debreu (1954). They demonstrate that perfect competition maximizes social welfare under these assumptions. This result may be used to derive the performance of a country as if this country maximizes its own welfare subject to constraints imposed by the transformation function. In autarky, country k maximizes its welfare subject to its own production possibilities. In this case, domestic consumption must necessarily be equal to domestic production, that is, $c_k{}^1 = q_k{}^1$, $c_k{}^2 = q_k{}^2$. The maximization problem can be solved by maximizing the Langrangian L^I:

$$L^I = V_k \left[c_k^1, c_k^2 \right] + \lambda^I T_k \left[q_k^1, q_k^2 \right] \quad \text{with } c_k^1 = q_k^1, \ c_k^2 = q_k^2. \quad (4.94)$$

The necessary conditions imply

$$\frac{\partial V_k(\cdot)/\partial c_k^1}{\partial V_k(\cdot)/\partial c_k^2} = \frac{\partial T_k(\cdot)/\partial q_k^1}{\partial T_k(\cdot)/\partial q_k^2}. \quad (4.95)$$

If country k is able to change goods from own production into goods for own consumption using an exchange rate p, the constraint that domestic consumption must coincide with domestic production is relaxed. p denotes the price ratio p_1/p_2 and mirrors the opportunities of country k for exchanges on the world market because p units of good 2 can be changed into one unit of good 1 and vice versa. Of course, every trading must be mutual so that the value of received goods equals the value of goods which were given away. This condition defines a budget constraint which requires that the sum of import values should be zero. Let imports (a negative sign of which indicates exports) be denoted by m. Then, the maximization problem can be solved by maximizing the Lagrangian L^{II} for any p:

$$L^{II} = V_k \left[q_k^1 + m_k^1, q_k^2 + m_k^2 \right] + \lambda^{II} T_k \left[q_k^1, q_k^2 \right] + \mu^{II} \left[p m_k^1 + m_k^2 \right]. \quad (4.96)$$

By definition, $q_k + m_k$ equals consumption. The necessary conditions for an interior solution are

$$\frac{\partial V_k(\cdot)/\partial c_k^1}{\partial V_k(\cdot)/\partial c_k^2} = p = \frac{\partial T_k(\cdot)/\partial q_k^1}{\partial T_k(\cdot)/\partial q_k^2}. \tag{4.97}$$

Equation (4.96) is maximized by determining production and import levels. Import levels of zero were optimal only if

$$\frac{\partial V_k\left[m_k^1 = 0\right]/\partial c_k^1}{\partial V_k\left[m_k^2 = 0\right]/\partial c_k^2} = \frac{\partial T_k[\cdot]/\partial q_k^1}{\partial T_k[\cdot]/\partial q_k^2} = p \tag{4.98}$$

holds. Only if the marginal rate of substitution and the marginal rate of transformation equalled p in autarky, would the solutions for autarky and trade be the same. In this case, the country would not import or export any good. In all other cases, however, the constraint implied by autarky, that is, $c_k = q_k$, is not met by the optimal import plans of country k. If the marginal rate of transformation under autarky falls short of (exceeds) p, the country can improve on the autarky outcome by restricting (increasing) the production of good 2 and increasing (restricting) the production of good 1 because the value of the increased production in one sector surmounts the value of the decreased production in the other sector.

As it is assumed that the world consists of only two countries, the direction of trade depends on the marginal rates of transformation of the two countries under autarky. Basically, the direction of trade was already determined by (4.92) and the resource endowments which determine (4.93). The relevant result of international trade is that a country specializes in the production of a good for which it has a relative price advantage – indicated by the marginal rate of transformation in autarky – because only relative prices, that is, exchange ratios, matter in this model. The utility function (4.92) influences this relative advantage because a country may have a strong preference for one good which raises its relative price in autarky. Hence, it may specialize in the production of the other good whose relative price is low in autarky. The relative resource endowment may also differ between both countries and a country specializes in the production of a good which employs intensively the resource with which the country is relatively well endowed compared to the other country. Both lines of reasoning are the explanations for relative price differences in autarky which were first developed by Heckscher and Ohlin. Alternatively, the relative productivity may differ between countries and a country specializes in the production of the good for which it has a relative productivity advantage. This is the explanation for relative price differences in autarky which was given by Ricardo.

Comparing the marginal rates of transformation which are equal to domestic relative prices in autarky shows the direction of trade. The volume of trade is determined by a world price p which implies consistent import and export plans of both countries. A world market equilibrium between country i and j is given if (4.99) holds.

$$\frac{\partial V_i\left[c_i^{1^*}, c_i^{2^*}\right] \Big/ \partial c_i^1}{\partial V_i\left[c_i^{1^*}, c_i^{2^*}\right] \Big/ \partial c_i^2} = \frac{\partial T_i\left[q_i^{1^*}, q_i^{2^*}\right] \Big/ \partial q_i^1}{\partial T_i\left[q_i^{1^*}, q_i^{2^*}\right] \Big/ \partial q_i^2} =$$

$$\frac{\partial V_j\left[c_j^{1^*}, c_j^{2^*}\right] \Big/ \partial c_j^1}{\partial V_j\left[c_j^{1^*}, c_j^{2^*}\right] \Big/ \partial c_j^2} = \frac{\partial T_j\left[q_j^{1^*}, q_j^{2^*}\right] \Big/ \partial q_j^1}{\partial T_j\left[q_j^{1^*}, q_j^{2^*}\right] \Big/ \partial q_j^2} = p,$$

$$c_i^{1^*} - q_i^{1^*} = q_j^{1^*} - c_j^{1^*}. \qquad (4.99)$$

Condition (4.99) guarantees that both countries' export and import plans are consistent because the total excess demand for both goods is zero. Note that the last condition of (4.99) is sufficient because the zero excess demand condition for good 2 follows from it together with the budget constraints. Equation (4.99) is also the result of world welfare maximization. World welfare maximization is the solution of maximizing the Langrangian

$$L^W = V_i\left[c_i^1, c_i^2\right] + V_j\left[c_j^1, c_j^2\right] + \lambda_i T_i\left[q_i^1, q_i^2\right] + \lambda_j T_j\left[q_j^1, q_j^2\right]$$
$$+ \mu^1\left[q_i^1 + q_j^1 - c_i^1 - c_j^1\right] + \mu^2\left[q_i^2 + q_j^2 - c_i^2 - c_j^2\right]. \qquad (4.100)$$

Equation (4.100) does not contain any budget constraint but specifies that the world excess demand must be zero. This condition is also implied by the budget constraints of both countries which (4.96) has employed to derive the optimal solution for one country. Thus, (4.99) is not just any trade equilibrium which improves on the autarky welfare of both countries but a trade equilibrium which maximizes world welfare.

Effects of a Unilateral Introduction of Trade Policies

The basic question to be asked when discussing trade policy is whether one country can improve on this outcome by unilaterally introducing trade barriers for the imports from the other country. Trade barriers like tariffs, import quotas and other restraints have a twofold impact:

- First, they decouple the domestic price from the world price because every trade barrier is likely to raise the prices of the import good, and the price rise must be carried by domestic consumers under perfectly competitive market structures.
- Second, the increase in the import good price lets the demand for the import good decrease and the domestic production of this good increase. The decreased consumption and the increased own production of this good may change the world price because the relative price of the import good decreases on the world market.

If a country is able to trigger the second effect, that is, a declining world market price, it will be shown to be able to improve on its free trade welfare by unilaterally introducing trade barriers. This effect results from the terms-of-trade effect: if a country is able to decrease the relative price of the import good, it is able to increase the relative price of its export good. The ratio of the export good price to the import good price defines the terms of trade if only two goods are assumed. An increase in the terms of trade increases the real income of a country because it receives more import goods for a given quantity of export goods when it improves its terms of trade. According to the usual terminology, a country is called large if it is able to alter its terms of trade by trade policies; it is called small if trade policy cannot alter the terms of trade.

The income effect can be explained by considering consumers and producers who maximize the consumers' utility subject to a budget constraint which contains the domestic prices. It is assumed that country k imports good 2 and exports good 1 under free trade. Every trade policy will therefore concern good 2. Additionally, the degree of trade policies, that is, the severity of trade restrictions, will be denoted by z so that a higher z indicates stricter import restrictions and $z = 0$ denotes no restrictions, that is, free trade. The domestic prices which are a function of the degree of trade policies introduced by country k are denoted by '$\wedge$'. Households and firms in a perfectly competitive environment behave as if they maximized the Langrangian (4.101),

$$L^{III} = V_k \left[c_k^1, c_k^2 \right] + \lambda^{III} T_k \left[q_k^1, q_k^2 \right] + \mu^{III} \left[\hat{p}(z) q_k^1 + q_k^2 - \hat{p}(z) c_k^1 - c_k^2 + y(z) \right]$$

(4.101)

for a given z. $y(z)$ denotes the income due to trade policies, for example the income arising from the imposition of a tariff. As households and firms only determine quantities to be produced and to be consumed, the necessary conditions for a given z result in

$$\frac{\partial V_k(\cdot)/\partial c_k^1}{\partial V_k(\cdot)/\partial c_k^2} = \hat{p}(z) = \frac{\partial T_k(\cdot)/\partial q_k^1}{\partial T_k(\cdot)/\partial q_k^2}.$$

(4.102)

Equations (4.102) and (4.97) coincide if $z = 0$ holds. Thus, free trade, that is, $z = 0$, is only optimal for country k if maximization of (4.101) including the country's trade policy implies $z = 0$. Therefore, free trade is the equilibrium in a one-shot game of international trade policies only if condition (4.103) holds:

$$\frac{\partial L^{III}}{\partial z}\left(z^* = 0 \right) = \mu^{III} \left[\frac{d\hat{p}}{dz}\left(z^* = 0 \right) \left[q_k^{1*} - c_k^{1*} \right] + \frac{dy}{dz}\left(z^* = 0 \right) \right] \leq 0.$$

(4.103)

z^* denotes the optimal trade policy. Equation (4.103) requires that free trade should qualify for an international trade policy equilibrium so that $z^* = 0$ maximizes the country's welfare given free trade policy of the other country. Condition (4.103) assumes that trade policies always aims at restricting imports and rules out that a country promotes imports by a negative z. Then, the derivative at $z^* = 0$ must be non-positive. This is a reasonable assumption since promoting imports would decrease the country's terms of trade.

In (4.101), the budget constraint was given as a function of the domestic price. In order to consider income effects in more detail, the relationship between the domestic price $\hat{p}$ and the world market price $\tilde{p}$ must be cleared:

$$\hat{p} := \frac{\tilde{p}}{1+z}, \quad \tilde{p} := \tilde{p}(z), \quad \frac{d\tilde{p}}{dz} \geq 0, \quad \tilde{p}(0) = p$$

$$\Rightarrow \frac{d\hat{p}}{dz} = -\frac{\tilde{p}}{(1+z)^2} + \frac{d\tilde{p}/dz}{1+z} \overset{!}{<} 0$$

(4.104)

Equation (4.104) specifies that the domestic price $\hat{p}$ is a function of the degree of domestic trade policies and that the domestic price decreases with the degree of trade policies compared to the world market price (recall that every p term denotes the price ratio p_1/p_2). In (4.104), the impact of trade policies on the domestic price is modelled as a tariff equivalent because a tariff of level z would change domestic prices exactly according to (4.104). The world market price is assumed not to be decreased by unilateral trade policies of country k. According to the usual terminology, a large country faces $d\tilde{p}/dz > 0$ and a small country faces $d\tilde{p}/dz = 0$. If $z = 0$, world price and domestic price are equal to p. The change of the domestic price as a response to trade policies is given in the last line of (4.104). It consists of a negative effect which is due to higher domestic prices and a positive effect which is due to lower world market prices. It is assumed that the second effect does not overcompensate the first effect.

The income effect of trade policies consists of two effects. First, the introduction of trade policies is assumed to raise the domestic income as importers have to either pay a tariff or pay for an import allowance, and these funds are distributed to domestic consumers as lump-sum payments. Second, the country increases its real income by improving its terms of trade. Definition (4.105) gives both effects for the free trade production and consumption levels:

$$y(z) := z\left[c_k^{2^*} - q_k^{2^*} \right] + \left[\tilde{p}(z) - \tilde{p}(0) \right]\left[q_k^{1^*} - c_k^{1^*} \right]. \tag{4.105}$$

Of course, $y(0)$ is zero for the free trade equilibrium. The first term gives the trade policy revenues and the second term gives the income increase which results from improved terms of trade. Definition (4.105) can be rewritten by the use of (4.106):

$$\tilde{p}q_k^{1^*} + q_k^{2^*} - \tilde{p}c_k^{1^*} - c_k^{2^*} = 0$$

$$\Rightarrow y(z) = \left[q_k^{1^*} - c_k^{1^*} \right]\left[\tilde{p}(z)z + \tilde{p}(z) - \tilde{p}(0) \right],$$

$$\frac{dy}{dz} = \left[q_k^{1^*} - c_k^1 \right]\left[\frac{d\tilde{p}}{dz}z + \tilde{p}(z) + \frac{d\tilde{p}}{dz} \right]. \tag{4.106}$$

Equation (4.106) observes that the budget constraint which country k has to meet is based on the world market price because the exchange possibilities of country k do not depend on the domestic price but on the terms of trade. This fact allows (4.105) to be rewriten only in terms of the export good 1. The last line gives the derivative of y with respect to z. This is the pure income effect which trade policies imply.

Definition (4.107) rearranges the LHS of the necessary condition for the optimality of $z^* = 0$ (see (4.103)):

$$\zeta(z) := -\frac{\tilde{p}(z)}{(1+z)^2} + \frac{d\tilde{p}(z)}{dz} z + \tilde{p}(z) + \left[1 + \frac{1}{1+z}\right] \frac{d\tilde{p}(z)}{dz}. \qquad (4.107)$$

ζ in (4.107) equals $[\partial L^{III}(z)/\partial z]/[\mu^{III}[q_k^{1*} - c_k^{1*}]]$ in (4.103) and uses (4.106). Since exporting good 1 means $q_k^{1*} - c_k^{1*} > 0$ and the shadow price μ^{III} of the budget constraint is clearly positive, the signs of $\partial L^{III}(z)/\partial z$ and ζ are identical. Thus, $z = 0$ is only optimal if $\zeta(z = 0)$ is not positive. $z = 0$ implies

$$\zeta(z = 0) = 2\frac{d\tilde{p}}{dz}(0) \geq 0. \qquad (4.108)$$

Equation (4.108) shows that $z = 0$ is only optimal for a small country for which $d\tilde{p}/dz = 0$, that is, a country which cannot improve on its terms of trade by adopting unilateral trade policies. In all other cases, every large country is able to improve on the free trade outcome by adopting unilateral trade policies. For $z \neq 0$, it can be explained easily that unilateral trade policies face a negative feedback which implies that the optimal degree of trade policies is not infinite. According to (4.102), consumers and producers determine consumption and production on the basis of domestic prices. Therefore, trade policies make firms produce more and households consume less of the import good. Consequently, the export level will decline with stricter import restrictions despite the increased terms of trade. When exports decline further and further with stricter trade policies, consumption of the import good approaches domestic production. Therefore, stricter trade policies imply consumption and production patterns which approach the autarky equilibrium so that the country has to sacrifice more and more of the benefits of free trade. The optimal unilateral trade policy balances the positive terms-of-trade effect and the negative change of both domestic consumption and production.

As both countries improve on their welfare by introducing unilateral trade policies, given no trade policy of the other country, and free trade maximizes world welfare, both countries enter a prisoners' dilemma situation in the standard cases of this model. If an international trade policy equilibrium exists, both countries must be expected to employ trade policy instruments. The existence of such a stable, non-cooperative equilibrium is not guaranteed, and if it exists it is also not clear whether countries end up both with a higher or a lower degree of trade policies compared to the optimal trade policies of each country (Johnson, 1953). If a stable, unique equilibrium exists, it is this non-cooperative, one-shot equilibrium which defines the utilities on which both countries may improve upon by strategic bargaining for a self-enforcing trade policy contract. It does not necessarily coincide with the autarky outcome which is a non-cooperative equilibrium if both countries' best responses to trade policies required the introduction of prohibitively high trade barriers (upon which no country could unilaterally improve). Since autarky can be reached by a finite degree of trade policies, the following simulation will assume that the autarky equilibrium is the non-cooperative equilibrium as well. This assumption is therefore not too artificial, and it simplifies the simulation because the non-cooperative payoffs are zero and bargaining gains are equal to payoffs.

The Chances of Free Trade

Because this section is exploring the possibilities for free trade as the result of a self-enforcing contract, it assumes that a non-cooperative international trade policy equilibrium exists for pure strategies which is unique and stable. This non-cooperative equilibrium defines the reference point upon which every self-enforcing contract should improve the outcome. Obviously, the set of self-enforcing contracts which improve this outcome also covers contracts which employ trade policy instruments. This section focuses now on the possibilities that strategic bargaining for a self-enforcing contract abolishes all trade barriers so that free trade is credibly and deliberately agreed upon. In this sense, free trade is the Pareto-optimal outcome as it maximizes both countries' total welfare. If trade policies were enforceable, free trade would be agreed upon, and bargaining gains would be split by the means of transfers.

If international trade contracts must be self-enforcing, (4.108) has demonstrated that every country may have an incentive to deviate from a free trade agreement by unilaterally introducing trade barriers. Thus, the short-run gains from a unilateral introduction of trade barriers should not exceed the long-run costs which defection implies through the reaction of the trade partner. Suppose that free trade gives a payoff of unity to each country. If free trade were the outcome of strategic bargaining for an enforceable trade contract, trade barriers would be abolished and transfers would be split according to

$$T = \frac{\ln \delta_j - \ln \delta_i}{\ln \delta_i + \ln \delta_j}. \tag{4.109}$$

Equation (4.109) follows from (4.80) under the assumption that $V_i = U_i = 1 + T$ and $V_j = U_j = 1 - T$. Under the free trade assumption, the relatively more patient country receives a higher share of the gains from free trade because it is able to appropriate the main part of the gains from trade. A self-enforcing contract which specifies free trade, however, has to fulfil credibility conditions as well:

$\forall k, l \in \{i, j\} \wedge \delta_k \geq \delta_l$:

$$\frac{1}{1-\delta_k}(1+T) - (bg+T) \geq 0,$$

$$\left[1 + \delta_k \frac{1-\delta_k^n}{1-\delta_k}\right](1+T) - (bg+T) + \delta_k \frac{1-\delta_k^n}{1-\delta_k} pg \geq 0,$$

$$\frac{1-\delta_k^n}{1-\delta_k} pg + \frac{\delta_k^n}{1-\delta_k}(1+T) \geq 0,$$

$$\frac{1}{1-\delta_l}(1-T) - bg \geq 0,$$

$$\left[1 + \delta_l \frac{1-\delta_l^n}{1-\delta_l}\right](1-T) - bg + \delta_l \frac{1-\delta_l^n}{1-\delta_l}(pg-T) \geq 0,$$

$$\frac{1-\delta_l^n}{1-\delta_l}(pg-T) + \frac{\delta_l^n}{1-\delta_l}(1-T) \geq 0, \quad bg > 0, \quad pg < 0. \tag{4.110}$$

Condition (4.110) specifies the three conditions which weak renegotiation-proofness imposes. It is assumed that both countries are symmetric. *bg* denotes the gains a country enjoys if it deviates from the free trade agreement, that is, the gains of introducing optimal trade policies. *pg* has a negative sign and denotes the punishment which a country has to bear when it invests in a resumption of cooperation by letting the other country introduce its optimal trade policies. (*pg* also indicates the loss for a country which suffers from defection of the other country.) (4.110) shows the conditions weak renegotiation-proofness imposes for a country k whose discount factor does not fall short of the discount factor of a country l. In this case, transfers according to (4.109) are not negative for k and are paid from country l to country k. As country k's discount factor gives it a comparably higher bargaining power and both countries are identical, country k never pays transfers.

Table 4.12   Trade policies for bg = 1.5 and pg = –0.5 and different discount factor combinations

δ_j \ δ_i	0.25	0.5	0.6	0,7	0.8	0.9	0.95
0.25	0	0	0	0	0	0	0
0.5	0	0	0	0	0	0	0
0.6	0	0	0	0	0	0	0
0.7	0	0	0	2	0	0	0
0.8	0	0	0	0	2	0	0
0.9	0	0	0	0	0	2	0
0.95	0	0	0	0	0	0	2

Note: Entries read: entry = 0: joint trade policies imply trade barriers, entry = $n > 0$: joint trade policies imply free trade for a punishment phase of n periods.

One expects free trade as the result of strategic bargaining for a self-enforcing trade contract if (4.110) holds for the transfer level specified by (4.109). If the transfer level specified by (4.109) violates one of the conditions of (4.110), no free trade will result from strategic bargaining but at least one country will employ trade barriers. Although high discount factors are known to improve the chances for cooperation, that is, the chances for the feasibility of free trade, relative differences between both countries are likely to prevent the free trade outcome even if it is feasible. The explanation is to be found in different bargaining powers: if a country has a comparably strong bargaining power, it would be able to realize transfers in the close neighbourhood of unity, given that free trade should result from such an agreement. This agreement, however, makes non-compliance of the donor country likely because its bargaining gains are smaller the more transfers it has to pay. Therefore, asymmetries with respect to discount factors may prevent a free trade agreement even if the feasibility of the free trade agreement is ensured by sufficiently high discount factors of both countries. Tables 4.12 and 4.13 demonstrate this effect.

Table 4.13 Trade policies for bg $= 1.25$ *and* pg $= -0.25$ *and different discount factor combinations*

δ_j \ δ_i	0.25	0.5	0.6	0.7	0.8	0.9	0.95
0.25	0	0	0	0	0	0	0
0.5	0	1	0	0	0	0	0
0.6	0	0	1	0	0	0	0
0.7	0	0	0	1	0	0	0
0.8	0	0	0	0	1	0	0
0.9	0	0	0	0	0	1	0
0.95	0	0	0	0	0	0	1

Note: Entries read: entry $= 0$: joint trade policies imply trade barriers, entry $= n > 0$: joint trade policies imply free trade for a punishment phase of n periods.

Tables 4.12 and 4.13 determine the chances of a free trade agreement by checking whether the bargaining-based free trade transfer level violates one of the compliance constraints given in (4.110). Table 4.12 uses a specification which makes non-compliance more attractive compared to Table 4.13. Both tables indicate that a self-enforcing trade contract which abolishes all trade barriers is only possible for identical discount factors. If discount factors differ, the self-enforcing contract will still employ trade barriers.

This result strengthens the strategic bargaining aspect. It emphasizes that transfers cannot automatically cure problems of potential non-compliance as they are subject to compliance risks as well. In the case of trade policies, a country is not interested in maximizing the world welfare but in maximizing its own welfare. If bargaining power does not match free trade, a country will employ trade barriers to exert its bargaining power. Therefore, one cannot translate the feasibility of free trade into the result that both countries agree upon a trade contract which abolishes all trade barriers. Such an agreement must qualify for the chosen contract by strategic bargaining, and the chances for free trade are the lower the more the bargaining power between both countries differ.

4.8 SUMMARY

This chapter has demonstrated both the relevance of transfers for compliance and strategic bargaining and the increased complexity of allowing transfers to support joint effort policies. Although twelve cases had to be distinguished when discussing strategic bargaining for a collective good with transfers, many salient results could be developed:

- Transfers must not be dealt with as enforceable if all other coordinated policies have to be self-enforcing.

  This requirement prevents effort and transfer policies from being separated from the division of bargaining gains. If transfers were enforceable, both agents could maximize the sum of individual bargaining gains subject to their compliance constraints. It is, however, unsatisfactory to assume enforceability of transfers but non-enforceability of joint effort policies. If the consistent assumption is made that every transfer policy must be self-enforcing as well, the division of bargaining gains and the allocation of efforts and transfers cannot be separated.

- Transfers which are subject to compliance risks as well double the number of compliance constraints compared to the case without transfers.

  This doubling is due to the different roles played by a donor and a receiver: if certain transfers in addition to a certain joint effort policy are agreed upon, the donor deviates by defecting from his/her effort policy and refuses the transfer and the receiver deviates by defecting from his/her effort policy but takes the transfer.

- The optimal policies for each agent do not specify the Pareto-optimal levels. The optimal policies may specify positive, zero or negative transfers.

  The Pareto-optimal levels specify the effort levels which maximize the sum of bargaining gains without any compliance constraint. If contracts were enforceable, the optimal policies would give all bargaining gains to the optimizing agent (and would leave the other agent without gains). This outcome would obviously violate a compliance constraint.

Therefore, the optimal policy of an agent does not employ the Pareto-optimal levels in a non-cooperative environment.

• The sign of an optimal policy transfer level depends on the discount factors and the benefit and cost parameters. If the Pareto-optimal levels give an agent negative bargaining gains without transfers because both his/her benefits and costs are low, the other agent's optimal policy specifies that the first agent receives transfers. Conversely, if costs and benefits are similar, the optimal policy of an agent specifies that he/she receives transfers.

This result mirrors the possibility that one agent has a seller attribute so that his/her marginal benefits and costs are low compared to the other agent. In this case, the benefits of paying this agent for a high effort level overcompensate the benefits of received transfers.

• Given that optimal policies specify non-zero transfers, concessions will be made by changing transfers and one agent's effort level until the Pareto-optimal levels are reached (if feasible). If the Pareto-optimal levels are reached, concessions will be made using transfers. Thus, concessions will not be based on changing all three variables but only on changing two or one of them, respectively. Zero transfers for an agent's optimal policy are possible and emphasize the results developed by Chapter 3 even if transfers are allowed. If zero transfers are specified by an optimal policy, concessions start with changing both agents' effort levels until one agent has reached his/her Pareto-optimal level. In this range of concessions, transfers are fixed on a zero level.

Section 4.4 has demonstrated that maximally two out of the three variables will be changed along the concession line. It has also shown that the concession policies modelled by Chapter 3 may still play a decisive role because transfers do not enter concession policies unless at least one agent has reached his/her Pareto-optimal level.

• Strategic bargaining may specify a self-enforcing contract which does lie on a compliance constraint curve even if the Pareto-optimal levels are feasible.

This result emphasizes again that the feasibility of the Pareto-optimal levels does not decide which outcome will be chosen. Instead, this outcome must be feasible and the result of strategic bargaining. The impact of this result has been demonstrated for trade policies in Section 4.7. The Pareto-optimal levels of trade policies specify that both countries abolish all trade barriers because free trade maximizes the sum of both countries' welfare. But each country is interested in not maximizing world welfare but its individual welfare subject to compliance of the other country and potential delay costs in strategic bargaining. If one country has a comparably strong bargaining position, it could seize a huge part of the gains from free trade of the other country if all trade barriers were to be abolished. This division of gains, however, makes non-compliance of the other country more likely the stronger is the bargaining power of this country. It was shown on the basis of two numerical examples that free trade is unlikely to qualify for a strategic bargaining solution if bargaining powers between countries differ.

- Feasibility constraints may restrict the concession line to ranges which do not allow transfers. Thus, even if transfers are allowed, they may play no role but the problem may be dominated by a concession line which was described in Chapter 3.

  If transfers can play a role, it is not guaranteed that they will be employed in every case. Thus, one should not conclude that an agreement which specifies no transfers although transfers are allowed is an irrational outcome. Instead, it may mirror the result of strategic bargaining between two partners who deliberately do not employ transfers.

- The impact of discount factor changes is ambiguous, too, if this discount factor defines the relevant range of the concession line.

  Along the concession line, compliance constraints exist which employ transfers. In this range, changing the discount factor of the agent who defines the compliance constraint under consideration produces both a bargaining and a compliance effect; the total impact of both is ambiguous. Thus, the ambiguity result is not restricted to the range of zero transfers but covers non-zero transfers as well.

- Comparing the distribution of bargaining gains in cases with and without transfers demonstrates that the role of self-enforcing transfers is more restricted than expected at first glance. If both agents have similar costs and benefits, a transfer- and bargaining-based self-enforcing contract may have different impacts: compared to the case without transfers, one agent may lose whereas the other agent gains. Transfer-based contracts, however, are likely to benefit both agents if they allow significant cost–benefit differentials between both agents to be exploited.

  This result stresses that transfers do not qualify for curing compliance problems but for exploiting cost and benefit differentials between two agents. They play a dominant role in strategic bargaining for self-enforcing contracts when they may be used to pay an agent who has the seller attribute. The simulation for environmental cooperation in Table 4.11, however, has also demonstrated that transfers play a minor role if one agent has a significantly low discount factor and his/her seller attribute cannot overcompensate for the compliance effect.

5. Bargaining between Buyer and Seller

5.1 INTRODUCTION: SELF-ENFORCING COOPERATION BETWEEN BUYER AND SELLER

The previous two chapters assumed a collective good problem which is defined by the assumption that individual efforts benefit both agents. Transactions between a buyer and a seller are not of that kind because every action taken unilaterally by one of them affects only the other agent. A buyer is an agent who enjoys a certain service or consumes a certain good which is produced or delivered and sold by the seller. The seller has to carry the production costs and wants to be compensated by transfers paid by the buyer. If transfers are impossible, the seller will not deliver because he/she only benefits from transfers.

Most of the applied bargaining literature deals with price determination between buyer and seller (see, for example, Hoel, 1986b) or between several buyers and sellers where third parties' actions are modelled as outside options for every agent (for an application to spatial competition, see Bester, 1989b). The strategic bargaining approach can determine the price paid for a certain good because the difference between the production costs of the seller and the utility of the buyer defines the bargaining gains. If the bargaining power is concentrated mainly on the buyer or the buyer has some 'good' outside options, the price will cover only a little more than production costs; if the bargaining power is concentrated mainly on the seller or the seller has some 'good' outside options, the price will be little less than the buyer's willingness to pay. Thus, price determination between buyer and seller is a classical application of the strategic bargaining approach.

If contracts between buyer and seller must be self-enforcing, repetition must guarantee that both agents keep the contract. One prominent application of self-enforcing contracts between buyer and seller is illegal contracts. Illegal contracts are not backed by any legal enforcement mechanism. For example, illegal labour contracts give both agents an option of non-compliance: the worker may take his/her wage without work input, the employer may refuse to pay the worker after work. Unless illegal enforcement institutions are replaced by legal enforcement, every illegal contract is subject to compliance risks. Other applications in this field are illegal transactions for drugs or for insider information in financial business.

However, illegal markets are often governed by a monopolistic position of the seller, for example on illegal drug markets. Offering an illegal product implies high risk costs which can often be carried only by a single supplier. Therefore, the distribution of bargaining power is often determined by the transaction costs of operating in an illegal market. If an illegal market is at least regionally dominated by a monopolist, the seller is able to set the conditions for transactions. This is the reason why transactions are based on rules which favour the seller. For example, a monopolistic employer may tear a banknote and give one part to the worker who receives the other part only in the case of agreed-upon work input. Any strategic bargaining model is therefore not applicable to illegal contracts when one agent has a monopolistic position. Strategic bargaining is restricted to really bilateral contracts for which the influence of third parties or of outside options is not too strong. For example, a speculator seeking for some insider information from a bank clerk may reflect bilateral trading better than drug deals or illegal labour contracts.

Contracts between buyer and seller which must be self-enforcing are not restricted to illegal contracts. Legal contracts are also subject to compliance risks if the transaction costs of enforcing the contract exceed the gains from enforcement. The set of legal contracts which must be self-enforcing depends on both the potential gains to be expected and the costs of bringing a broken contract into court. For example, transaction costs arise because the accusing party may have to carry the costs of a legal adviser and court costs, may have to prove non-compliance of the other party and may have to realize significant delay between contract breach and compensation (which depends on the individual impatience as well). These costs depend on the institutional setting which divides the set of legal contracts into those which must be self-enforcing and those which are enforceable. Contracts will always be designed as self-enforcing if the compensation to be expected by bringing breaches into court is so low that the punishment plan according to weak renegotiation-proofness ensures earlier compensation for non-compliance.

Another field of self-enforcing contracts between buyer and seller is contracts between two parties in different countries. Unless binding agreements guarantee identical treatment of domestic and foreign agents bringing a contract breach into court, all agreements among agents belonging to different sovereign countries are subject to compliance risks. One application is international environmental policies which aim at restricting upstream–downstream problems of pollution. In this case, the countries themselves are the agents under consideration. Upstream–downstream pollution defines no collective bad because the upstream country pollutes only the downstream country by an uncoordinated resource management but does not suffer from this policy. This is a typical buyer–seller case when the upstream country sells an ambitious national environmental policy to the downstream country. A sovereign upstream country may opt to take the transfer but not to revise national resource use; a sovereign downstream country may refuse to pay for a revised national resource use of the upstream country.

Other applications are all variants of international project management. International project management involves at least two parties from different countries. One party plans a national project to be realized with the assistance of all other parties. This party is typically the buyer of project support which is sold by the other party. An investment which is financed by a foreign bank reflects one kind of project support and directly demonstrates the relevance of non-compliance options in an international setting. The debt crisis has impressively shown that international contracts must be self-enforcing because every sovereign country is able to repudiate debt services (for the literature on sovereign debt, see Atkeson, 1991; Bulow and Rogoff, 1989; Mohr, 1991). Although history gives only examples of non-compliance of the debtor, non-compliance of the creditor is also imaginable if the debt contract is based on some securities which can be seized unilaterally by the creditor.

International resource exploitation under expropriation risks gives another example of international project management (see Thomas and Worrall, 1994). In this case, a company erects facilities in a foreign country, say, in order to produce oil. This company can be assumed to be able to produce oil very efficiently whereas the foreign country is not able to do so alone. Technology transfers and management skills play a major role in subcontracting national resource exploitation to foreign companies. The company pays a royalty to the country to be allowed to drill for oil. The company may opt to produce oil without paying royalties (and then to leave the country), and the country may expropriate the company in order to realize the short-run gains of expropriating the oil production (which makes the foreign company leave the country). In this case, the foreign country and the company are the seller and the buyer of the right to produce oil, respectively.

The organization of this chapter follows the structure of the previous chapters. Section 5.2 develops the compliance constraints, the optimal policies and the concession line for the buyer–seller model. As benefits for the buyer and costs for the seller are given other functional forms, feasibility aspects for the optimal policies and the Pareto-optimal level will be discussed more explicitly. Section 5.3 gives the strategic bargaining solution and discusses the impacts of discount factor changes. Additionally, it contains some simulations which compare prices implied by self-enforcing contracts with prices implied by enforceable contracts. Section 5.4 concludes and summarizes this chapter.

5.2 COMPLIANCE CONSTRAINTS, OPTIMAL POLICIES AND THE CONCESSION LINE

This chapter assumes also specific utility functions. After introducing the utility functions of the buyer and the seller, a short digression about how bargaining gains would be divided if contracts were enforceable follows. This discussion compares price determination between enforceable and self-enforcing contracts. Then, this section will continue by determining both agents' compliance constraints. Instead of repeating Chapter 3's discussion on optimal policies, it first assumes that both agents' optimal policies differ and second introduces a bargaining parameter which describes strategic bargaining solutions on one agent's compliance constraint. Finally, it discusses feasibility aspects for the optimal policies and the Pareto-optimal level.

This chapter assumes that the production and consumption of a good X produce linear benefits for the buyer and quadratic costs for the seller:

$$U_b = \alpha_b X - T, \quad U_s = T - \frac{\gamma_s}{2} X^2 \tag{5.1}$$

The subscripts b and s denote the buyer and the seller, respectively. T is the transfer paid for delivery of X quantities of the good. In a non-cooperative setting of a one-shot game, the buyer will not buy and the seller will not sell: the buyer does not pay anything for nothing and the seller does not produce anything for zero transfers. Hence, the non-cooperative utilities are zero. This result simplifies the determination of the strategic bargaining solution and allows an explicit solution for a strategic bargaining parameter to be computed. In a cooperative setting, both agents would unanimously agree upon a delivery $X = \alpha_b/\gamma_s$ which maximizes total utility. Bargaining would concern only transfers. As the Pareto-optimal level would not be subject to bargaining, bargaining for transfers would translate easily into bargaining for the price per unit, $p = T/X$. The minimum and maximum transfers can be determined by setting U_b and U_s equal to zero for the Pareto-optimal levels:

$$T_{\max} = \frac{\alpha_b^2}{\gamma_s}, \quad T_{\min} = \frac{\alpha_b^2}{2\gamma_s} \tag{5.2}$$

$T_{\max}$ ($T_{\min}$) gives the buyer (seller) no bargaining gains but the seller (buyer) all the bargaining gains. Because bargaining would be concentrated on transfers, $dU_b = -dU_s$ would hold and developing the strategic bargaining solution would be straightforward:

$$-\frac{dU_b}{dU_s}\frac{U_s}{U_b} = \frac{\ln\delta_b}{\ln\delta_s} \Leftrightarrow \frac{T-(\gamma_s/2)X^2\alpha}{\alpha_b X - T} = \frac{\ln\delta_b}{\ln\delta_s}$$

$$\Leftrightarrow T = \frac{\alpha_b^2}{2\gamma_s}\frac{2\ln\delta_b+\ln\delta_s}{\ln\delta_b+\ln\delta_s}, \quad p = \frac{\alpha_b}{2}\frac{2\ln\delta_b+\ln\delta_s}{\ln\delta_b+\ln\delta_s} \tag{5.3}$$

Equation (5.3) gives the strategic bargaining solution for the transfer level and the price per unit if contracts were enforceable. It demonstrates that price determination in a bilateral setting does not depend on the production and consumption level because enforceable contracts always specify the level which maximizes total utility. This digression on enforceable contracts should therefore have demonstrated that price setting is not a behavioural assumption but the strategic variable. This result carries over to self-enforcing contracts. The next section will use (5.3) in order to compare prices which result from bargaining for a self-enforcing contract with prices which were the result of a corresponding enforceable contract.

Coming back to self-enforcing contracts, a self-enforcing contract between a buyer and a seller has to meet three conditions for both buyer and seller. The non-compliance option means that the buyer consumes the goods produced by the seller but does not pay him/her. For the seller, it means he/she takes the transfers from the buyer but does not produce the good. In the first case, the seller carries the production costs without any compensation by the buyer; in the second case, the buyer pays transfers for nothing.

Starting with the buyer's compliance constraints, the first condition requires that the buyer's benefits from always fulfilling the contract must not fall short of the current benefits from breaking it and no cooperation in all following periods:

$$\frac{\alpha_b X - T}{1-\delta_b} \geq \alpha_b X. \tag{5.4}$$

Weak renegotiation-proofness was shown to imply two conditions which define *ex ante* and *ex post* compliance, and which fulfil (5.4). As in Section 4.2, this section assumes one period of punishment. This assumption is justified by observing that all three conditions also coincide for the model of this chapter. *Ex ante* compliance requires

$$(1+\delta_b)(\alpha_b X - T) \geq \alpha_b X - \delta_b T; \tag{5.5}$$

ex post compliance requires

$$-T + \frac{\delta_b}{1-\delta_b}(\alpha_b X - T) \geq 0. \tag{5.6}$$

Equations (5.5) and (5.6) indicate that punishment of the buyer means that he/she has to pay the agreed-upon transfers without consuming the good. Equation (5.5) requires that the discounted gains from the self-enforcing contract should not fall short of breaking the contract in one period and resuming cooperation in the next period. Equation (5.6) requires that investing in a resumption of cooperation should be preferred by a non-compliant buyer.

Equations (5.4), (5.5) and (5.6) coincide because of the orthogonality of the reaction functions and thereby justify the assumption of one-period punishments. The agreed-upon transfer and production levels should fulfil

$$T \leq \delta_b \alpha_b X. \tag{5.7}$$

Equation (5.7) collects (5.4), (5.5) and (5.6) and requires that the agreed-upon transfers should not exceed a certain level because otherwise the buyer would consume the good without paying the seller. For the seller's compliance constraints, the three conditions imposed by weak renegotiation-proofness are

$$\frac{1}{1-\delta_s}\left[T - \frac{\gamma_s}{2}X^2\right] \geq T, \tag{5.8}$$

$$[1+\delta_s]\left[T - \frac{\gamma_s}{2}X^2\right] \geq T - \delta_s \frac{\gamma_s}{2}X^2, \tag{5.9}$$

$$-\frac{\gamma_s}{2}X^2 + \frac{\delta_s}{1-\delta_s}\left[T - \frac{\gamma_s}{2}X^2\right] \geq 0. \tag{5.10}$$

Equation (5.8) gives the condition that keeping the self-enforcing contract should benefit the seller more than breaking it in one period and no cooperation in all future periods. Equations (5.9) and (5.10) give the conditions for *ex ante* and *ex post* compliance, respectively, for a one-period punishment, and they imply (5.8). Both conditions indicate that punishment of the seller means that he/she has to produce the agreed-upon quantity and to carry the resulting costs without receiving transfers from the buyer.

Equations (5.8), (5.9) and (5.10) coincide, too, and thereby justify the assumption of one-period punishments as well. The agreed-upon transfer and production levels should fulfil

$$T \geq \frac{1}{\delta_s} \frac{\gamma_s}{2} X^2. \tag{5.11}$$

Equation (5.11) collects (5.8), (5.9) and (5.10) and requires that the agreed-upon transfers should not fall short of a certain level because otherwise the seller would take the transfer without producing the good.

As has been announced at the beginning, this section will ignore feasibility constraints in a first step when discussing optimal policies and the concession line. Ignoring feasibility constraints with respect to optimal policies means assuming that the optimal policy for one agent which is determined along the compliance constraint of the other agent is not made infeasible by the compliance constraint of the agent who is supposed to pursue his/her optimal policy. In this case, the optimal policy for the buyer is the solution of maximizing his/her utility subject to the seller's compliance constraint (5.11). Since T decreases the buyer's utility, one may use directly the limit of (5.11):

$$\max_{X} \{\alpha_b X - T\} \quad \text{s.t.} \ T = \frac{1}{\delta_s} \frac{\gamma_s}{2} X^2$$

$$\Rightarrow X_b^* = \delta_s \frac{\alpha_b}{\gamma_s}, \ T_b^* = \delta_s \frac{\alpha_b^2}{2\gamma_s}. \tag{5.12}$$

The star denotes optimality for the agent who is denoted by the subscript. The optimal solution can also be developed in a diagram. Since $dT = \gamma_s X/\delta_s$ for the frontier of (5.11) and $dU_i = 0$ implies $dT/dX = \alpha_b$, the critical transfer level is a quadratic function of X and the iso-utility curves of the buyer are linear as shown in Figure 5.1.

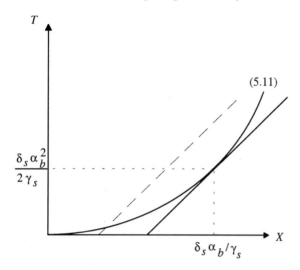

Figure 5.1   Optimal policy for the buyer

The unbroken line gives the iso-utility curve for the buyer which is tangential to the limit of (5.11). This point defines the buyer's optimal policy. The broken line depicts another iso-utility curve which gives the buyer a lower utility. The buyer's utility is increased if the iso-utility curves are shifted rightwards since an increase in X for a given T increases his/her utility.

The optimal policy for the seller is the solution of maximizing his/her utility subject to the buyer's compliance constraint (5.7). Since T benefits the seller, one may use directly the limit of (5.7):

$$\max_{X}\left\{T - \frac{\gamma_s}{2}X^2\right\} \quad \text{s.t.} \ T = \delta_b\alpha_b X$$

$$\Rightarrow X_s^* = \delta_b\frac{\alpha_b}{\gamma_s}, \ T_s^* = \delta_b^2\frac{\alpha_b^2}{\gamma_s}. \tag{5.13}$$

Figure 5.2 depicts the optimal solution. Since $dT = \delta_b\alpha_b dX$ for the frontier of (5.7) and $dU_i = 0$ implies $dT/dX = \gamma_s X$, the critical transfer level is a linear function of X and the iso-utility curves of the buyer are quadratic.

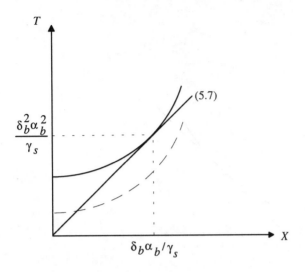

Figure 5.2   Optimal policy for the seller

The unbroken line gives the iso-utility curve for the seller which is tangential to the limit of (5.7). This point defines the seller's optimal policy. The broken curve depicts another iso-utility curve which gives the seller a lower utility. The seller's utility is increased if the iso-utility curves are shifted upwards since an increase in T for a given X increases his/her utility.

Note that both agents' optimal policies imply a production and consumption level which falls short of the Pareto-optimal level α_b/γ_s.

If both agents' compliance constraints allow the Pareto-optimal level to be realized, determining the concession line is straightforward. The buyer makes concessions away from his/her optimal policy along the seller's compliance constraint until the Pareto-optimal level is reached. The reason for this concession policy should be clear after the discussion of concession policies in Chapter 4. When the Pareto-optimal level is reached, any change in production does not make any sense and concessions will be exclusively transfer based.

Concessions along the seller's compliance constraint can be defined by the bargaining parameter Δ_s:

$$\delta_s \leq \Delta_s \leq 1: \quad X = \Delta_s \frac{\alpha_b}{\gamma_s}, \quad T = \frac{1}{\delta_s} \frac{\gamma_s}{2} X^2 = \frac{\Delta_s^2}{\delta_s} \frac{\alpha_b^2}{2\gamma_s} \tag{5.14}$$

Equation (5.14) allows both agents' utilities for concessions along the seller's compliance constraint to be determined as functions of the bargaining parameter Δ_s:

$$U_b(\cdot) = \Delta_s \frac{\alpha_b^2}{2\gamma_s}\left[2 - \frac{\Delta_s}{\delta_s}\right],$$

$$U_s(\cdot) = \Delta_s^2 \frac{\alpha_b^2}{2\gamma_s}\left[\frac{1}{\delta_s} - 1\right],$$

$$\frac{\partial U_b}{\partial \Delta_s} = \frac{\alpha_b^2}{\gamma_s}\left[1 - \frac{\Delta_s}{\delta_s}\right] \le 0,$$

$$\frac{\partial U_s}{\partial \Delta_s} = \Delta_s \frac{\alpha_b^2}{\gamma_s}\left[\frac{1}{\delta_s} - 1\right] > 0 \quad \text{for } \delta_s < 1. \tag{5.15}$$

Similarly, concessions along the buyer's compliance constraint can be defined by the bargaining parameter Δ_b,

$$\delta_b \le \Delta_b \le 1: \ X = \Delta_b \frac{\alpha_b}{\gamma_s}, \ T = \delta_b \alpha_b X = \delta_b \Delta_b \frac{\alpha_b^2}{\gamma_s}. \tag{5.16}$$

Equation (5.16) allows the determination of both agents' utilities for concessions along the buyer's compliance constraint as a function of the bargaining parameter Δ_b:

$$U_b(\cdot) = \Delta_b \frac{\alpha_b^2}{\gamma_s}(1 - \delta_b),$$

$$U_s(\cdot) = \Delta_b \frac{\alpha_b^2}{2\gamma_s}(2\delta_b - \Delta_b),$$

$$\frac{\partial U_b}{\partial \Delta_b} = \frac{\alpha_b^2}{\gamma_s}(1 - \delta_b) > 0 \quad \text{for } \delta_s < 1,$$

$$\frac{\partial U_s}{\partial \Delta_b} = \frac{\alpha_b^2}{\gamma_s}(\delta_b - \Delta_b) \le 0. \tag{5.17}$$

Two results which (5.15) and (5.17) show are noteworthy. First, (5.15) and (5.17) also demonstrate that the best policy for one agent still gives the other agent positive bargaining gains. Second, both agents make concessions along the other agent's compliance constraint by increasing the production level as well as the transfer level.

If feasibility constraints do not hinder purely transfer-based bargaining, the bargaining parameter τ defines both agents' utilities:

$$\frac{1}{\delta_s}\frac{\alpha_b^2}{2\gamma_s} \leq \tau \leq \delta_b \frac{\alpha_b^2}{\gamma_s},$$

$$U_b = \frac{\alpha_b^2}{\gamma_s} - \tau, \quad U_s = \tau - \frac{\alpha_b^2}{2\gamma_s},$$

$$\frac{dU_b}{d\tau} = -\frac{dU_s}{d\tau} = -1. \tag{5.18}$$

Equation (5.18) follows directly from the restrictions (5.7) and (5.11) for the transfer levels and from $X = \alpha_b/\gamma_s$ for the Pareto-optimal level. If feasibility constraints do not bind, (5.14), (5.16) and (5.18) describe the three ranges which together define the concession line.

The first line of (5.18) can be used to discuss the feasibility of the Pareto-optimal level α_b/γ_s. Rearranging terms shows that

$$\delta_b\delta_s \geq \frac{1}{2} \tag{5.19}$$

must hold to ensure that the set of feasible parameters τ is not empty. Figure 5.3 combines Figures 5.1 and 5.2 to show condition (5.19).

Figure 5.3 shows both critical transfer levels. Because all *T–X* combinations have to be found in the inner lens or on the boundaries, the intersection gives the highest feasible *T–X* combination. (Since (5.11) is quadratic with a zero slope at the zero transfer level and (5.7) is linear, Figure 5.3 also shows that there is always scope for improvement in this setting.) Equalizing the frontiers of (5.7) and (5.11) gives $X = 2\delta_b\delta_s\alpha_b/\gamma_s$ which must not fall short of α_b/γ_s to allow the Pareto-optimal level to be realized. This condition implies (5.19).

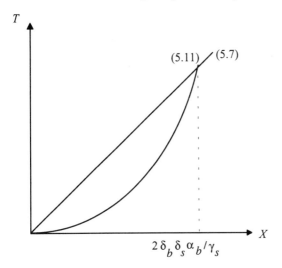

$$2 \delta_b \delta_s \alpha_b / \gamma_s$$

Figure 5.3   Feasibility of the Pareto-optimal level

Condition (5.19) is more demanding than the condition for the feasibility of a point on the contract curve of Chapter 3. It can be explained by reconsidering the implications of non-compliance in both models. In Chapter 3, non-compliance also meant a loss of own benefits because decreasing own efforts decreases the public good service. In this model, non-compliance does not mean sacrificing own benefits: if one agent breaks the contract, he/she does not lose anything. Therefore, the agents' discount factors are more important because there is no partially countervailing utility loss implied by non-compliance.

If (5.19) does not hold, the Pareto-optimal level is infeasible and the concession line consists of the feasible part which (5.14) and (5.16) describe. Equations (5.14) and (5.16), however, have assumed that one agent's optimal policy is not made infeasible by his/her own compliance constraint because they have developed optimal policies by assuming that only the other agent's compliance constraint is binding. If both optimal policies coincide so that both agents cannot be made better off by any agreement other than the one which specifies the intersection of both compliance constraints, the line of reasoning is quite similar to that in Chapter 3 and will therefore not be discussed explicitly in this section.

In order to make the optimal policy (5.12) of the buyer feasible, his/her utility must not fall short of the utility which he/she enjoys in the case that the seller pursues his/her optimal policy:

$$\delta_s \frac{\alpha_b^2}{2\gamma_s} \geq \delta_b \frac{\alpha_b^2}{\gamma_s}(1-\delta_b) \Leftrightarrow \delta_s \geq 2\delta_b(1-\delta_b). \tag{5.20}$$

Figure 5.4 depicts the case that (5.20) is violated.

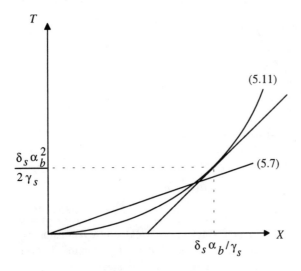

Figure 5.4   Infeasibility of the buyer's optimal policy

In Figure 5.4, (5.7) does not allow the optimal policy (5.12) to be realized because (5.12) would lead to non-compliance of the buyer.

The opposite holds for the optimal policy of the seller: his/her utility from (5.13) must not fall short of the utility he/she enjoys if the buyer realizes his/her optimal policy:

$$\delta_b^2 \frac{\alpha_b^2}{2\gamma_s} \geq \delta_s \frac{\alpha_b^2}{2\gamma_s}(1-\delta_s) \Leftrightarrow \delta_b \geq \sqrt{\delta_s(1-\delta_s)}. \tag{5.21}$$

Figure 5.5 depicts the case that (5.21) is violated. In Figure 5.5, (5.11) does not allow the optimal policy (5.13) to be realized because (5.13) would lead to non-compliance of the seller.

If (5.20) does not hold, the buyer's optimal policy (5.12) is not feasible; if (5.21) does not hold, the seller's optimal policy (5.13) is not feasible. Both conditions determine a lower bound for the one agent's discount factor which depends on the other agent's discount factor. The higher the one agent's discount factor is, the higher is the production level induced by the optimal policy according to (5.12) or (5.13) for the other agent and the more unlikely is infeasibility of (5.12) or (5.13), respectively.

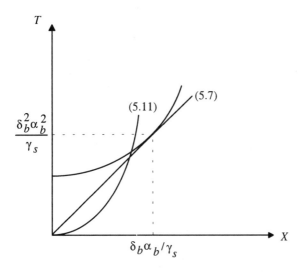

Figure 5.5   Infeasibility of the seller's optimal policy

Conditions (5.20) and (5.21) are not symmetric. Condition (5.20) can be rewritten using the functional Ψ_s:

$$\psi_s(\delta) = 2\delta(1-\delta),$$

$$\frac{d\psi_s}{d\delta} = 2(1-2\delta), \quad \frac{d^2\psi}{d\delta^2} = -4 < 0, \quad \Rightarrow \delta^{\max} = 0.5. \tag{5.22}$$

Equation (5.22) gives the minimum discount factor ψ_s of the seller as a function of the buyer's discount factor which is denoted by δ. Similarly, (5.21) gives

$$\psi_b(\delta) = \sqrt{\delta(1-\delta)},$$

$$\frac{d\psi_b}{d\delta} = \frac{1}{2}\left(\delta-\delta^2\right)^{-1/2}(1-2\delta),$$

$$\frac{d^2\psi_b}{d\delta^2} = -\frac{1}{4}\left(\delta-\delta^2\right)^{-3/2}(1-2\delta)^2 -\left(\delta-\delta^2\right)^{-1/2} < 0 \text{ for } 0 < \delta < 1,$$

$$\Rightarrow \delta^{\max} = 0.5. \tag{5.23}$$

Equation (5.23) gives the minimum discount factor Ψ_b of the buyer as a function of the seller's discount factor which is denoted by δ, too. Using δ in both functions allows both conditions to be compared and the impact of a given δ on Ψ_b and Ψ_s to be determined.

Comparing (5.22) and (5.23) shows that both functions have the same maximum $\delta = 0.5$. Equation (5.24) proves that this maximum gives the identity of Ψ_b and Ψ_s but Ψ_s falls short of Ψ_b for all other δ strictly between 0 and 1:

$$\psi_s(\delta) < \psi_b(\delta)$$
$$\Leftrightarrow 2\delta(1-\delta) < \sqrt{\delta(1-\delta)} \Leftrightarrow (\delta-0.5)^2 > 0 \Leftrightarrow \delta < 0.5 \vee \delta > 0.5. \quad (5.24)$$

Figure 5.6 shows the shape of both functions.

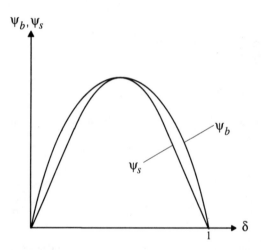

Figure 5.6   Feasibility of optimal policies in the buyer–seller case

Figure 5.6 and (5.24) demonstrate that the constraint which defines the minimum level of the buyer's discount factor is more demanding than the constraint which defines the minimum level of the seller's discount factor. *Ceteris paribus*, the buyer must be more patient than the seller in order to guarantee feasibility of optimal policies. This result is not due to the different role buyer and seller play but originates from the different assumptions made for the benefit function of the buyer and the cost function of the seller. As the seller faces quadratic production costs, his/her disutility from production increases overproportionally with production whereas the buyer's marginal utility is constant. Therefore, the constraint which defines the feasibility of the seller's optimal policy is more demanding. The opposite result would follow if (5.1) assumed constant marginal costs and marginally decreasing benefits. Thus, Figure 5.6 and (5.24) demonstrate that the different importance of both agents' discount factors also depends strongly on the assumed shape of the utility functions.

5.3 STRATEGIC BARGAINING SOLUTIONS AND THE IMPACT OF DISCOUNT FACTOR CHANGES

The previous section has described the concession line and has discussed the role of feasibility constraints. This section will develop the strategic bargaining solution because the outcome depends not only on feasibility aspects but on the relative impatience of both agents as well. This section will start by presenting the strategic bargaining solutions for the different ranges which define the concession line. Since $U_i = V_i$ and $U_j = V_j$, simple explicit expressions for both Δ_s and Δ_b can be given. Then, the impact of discount factor changes will be discussed. As the impact is quite obvious for ranges of the concession line which are not defined by the discount factor which is changed, this section concentrates on discount factor changes which alter the compliance constraint. It will show that the ambiguity result also holds for the buyer–seller case. Finally, this section will present some simulations. These simulations will also compare prices per unit between self-enforcing and enforceable contracts.

If the strategic bargaining solution is to be found along the seller's compliance constraint, the bargaining parameter Δ_s can be determined which equalizes $-\ln\delta_b/\ln\delta_s$ and ρ. According to (5.15),

$$\rho = \frac{dU_b}{dU_s}\frac{U_s}{U_b} = \frac{1-(\Delta_s/\delta_s)}{2-(\Delta_s/\delta_s)} = -\frac{\ln\delta_b}{\ln\delta_s}$$

$$\Leftrightarrow \Delta_s = \delta_s\frac{2\ln\delta_b+\ln\delta_s}{\ln\delta_b+\ln\delta_s} \Rightarrow X = \delta_s\frac{\alpha_b}{\gamma_s}\frac{2\ln\delta_b+\ln\delta_s}{\ln\delta_b+\ln\delta_s}$$

$$\Rightarrow p = \frac{T}{X} = \frac{\alpha_b}{2}\frac{2\ln\delta_b+\ln\delta_s}{\ln\delta_b+\ln\delta_s} \tag{5.25}$$

mirrors the strategic bargaining solution for the strategic bargaining parameter, the output level and the price per unit. Similarly, if the strategic bargaining solution is to be found along the buyer's compliance constraint, (5.17) gives

$$\rho = \frac{dU_b}{dU_s} \frac{U_s}{U_b} = \frac{2\delta_b - \Delta_b}{2(\delta_b - \Delta_b)} = -\frac{\ln \delta_b}{\ln \delta_s}$$

$$\Leftrightarrow \Delta_b = 2\delta_b \frac{\ln \delta_b + \ln \delta_s}{2\ln \delta_b + \ln \delta_s} \Rightarrow X = 2\delta_b \frac{\alpha_b}{\gamma_s} \frac{\ln \delta_b + \ln \delta_s}{2\ln \delta_b + \ln \delta_s}$$

$$\Rightarrow p = \frac{T}{X} = \delta_b \alpha_b = \text{const.} \tag{5.26}$$

Equation (5.26) reveals that the per unit price is identical for all possible strategic bargaining solutions along the buyer's compliance constraint. Concessions along the buyer's compliance constraint do not change the ratio between transfer and output level but, as the seller faces increasing costs, increasing the output and transfer level proportionally increases the seller's cost burden.

If δ_i $\delta_j \geq 0.5$ holds, the strategic bargaining solution may also specify the Pareto-optimal level and bargaining may be restricted to the transfer level. According to (5.18), the strategic bargaining solution is determined by (5.3) as this case coincides with bargaining for enforceable contracts. For this range, the bargaining parameter τ equals T in (5.3):

$$\tau = \frac{\left(\alpha_b^2/2\gamma_s\right)\left(2\ln \delta_b + \ln \delta_s\right)}{\ln \delta_b + \ln \delta_s},$$

$$X = \frac{\alpha_b}{\gamma_s} = \text{const.}, \quad p = \frac{\tau}{X} = \frac{\alpha_b}{2} \frac{2\ln \delta_b + \ln \delta_s}{\ln \delta_b + \ln \delta_s}. \tag{5.27}$$

It can be derived directly from (5.25), (5.26) and (5.27) that ρ is increased continuously from $-\infty$ for the optimal policies of the seller (see (5.26) for $\Delta_b \to \delta_b$) to 0 (see (5.25) for $\Delta_s \to \delta_s$) for the optimal policies of the buyer. Hence, uniqueness is guaranteed and feasibility of the points where the policies switch plays the same role as in Chapter 4.

Before turning to some simulations, the impact of discount factor changes will be addressed. Obviously, every change in a discount factor has an unambiguous effect on the agents' utilities unless this discount factor defines the relevant part of the concession line. If the discount factor which does not enter the relevant part of the concession line is changed, the incentives for compliance are not changed and the variation affects only $-\ln\delta_b/\ln\delta_s$. In this case, only the bargaining effect applies and increases (decreases) the utility of the agent whose relative patience is increased (decreased). This result holds for all variations if the bargaining parameter τ applies (α_b/γ_s remains unchanged), for variations of δ_b if the bargaining parameter Δ_s applies (δ_b is no argument in (5.15)) and for variations of δ_s if the bargaining parameter Δ_b applies (δ_s is no argument in (5.17)). The compliance effect plays a role in the remaining two cases. The first case considers variations of δ_s for solutions in the range of the seller's compliance constraint. The first line of (5.25) can be written as an implicit function:

$$F(\Delta_s, \delta_s) = \frac{\ln \delta_b}{\ln \delta_s} + \frac{1 - (\Delta_s / \delta_s)}{2 - (\Delta_s / \delta_s)} = 0,$$

$$\frac{\partial F}{\partial \delta_s} = -\frac{1}{\delta_s} \frac{\ln \delta_b}{(\ln \delta_s)^2} + \frac{\Delta_s}{\delta_s^2 [2 - (\Delta_s / \delta_s)]^2} > 0,$$

$$\frac{\partial F}{\partial \Delta_s} = -\frac{1}{\delta_s [2 - (\Delta_s / \delta_s)]} < 0,$$

$$\Rightarrow \left. \frac{d\Delta_s}{d\delta_s} \right|_{d\delta_b = 0} > 0. \tag{5.28}$$

Interpreting (5.15) as functions of both the bargaining parameter Δ_s and the discount factor δ_s (that is, keeping δ_b constant), total differentiation of both utility functions gives

$$\left. \frac{dU_b}{d\delta_s} \right|_{d\delta_b = 0} = \frac{\partial U_b}{\partial \delta_s} + \frac{\partial U_b}{\partial \Delta_s} \left. \frac{d\Delta_s}{d\delta_s} \right|_{d\delta_b = 0},$$

$$\left. \frac{dU_s}{d\delta_s} \right|_{d\delta_b = 0} = \frac{\partial U_s}{\partial \delta_s} + \frac{\partial U_s}{\partial \Delta_s} \left. \frac{d\Delta_s}{d\delta_s} \right|_{d\delta_b = 0}. \tag{5.29}$$

Both total differentials are ambiguous in sign: $\partial U_b / \partial \delta_s$ is positive but $\partial U_b / \partial \Delta_s \, d\Delta_s / d\delta_s$ is negative, $\partial U_s / \partial \delta_s$ is negative but $\partial U_s / \partial \Delta_s \, d\Delta_s / d\delta_s$ is positive. The ambiguity cannot be resolved by giving (5.29) explicitly because logarithmic terms enter both total differentials (see (5.28)) which prevent an unambiguous sign from being determined.

This ambiguity result also holds for the second case which considers variations of δ_b for solutions in the range of the buyer's constraint. Equation (5.30) gives the first line of (5.26) as an implicit function:

$$G(\Delta_b, \delta_b) = \frac{\ln \delta_b}{\ln \delta_s} + \frac{2\delta_b - \Delta_b}{2(\delta_b - \Delta_b)} = 0,$$

$$\frac{\partial G}{\partial \delta_b} = \frac{1}{\delta_b \ln \delta_s} - \frac{1}{2} \frac{\Delta_b}{(\delta_b - \Delta_b)^2} < 0,$$

$$\frac{\partial G}{\partial \Delta_b} = \frac{1}{2} \frac{\delta_b}{(\delta_b - \Delta_b)^2} > 0,$$

$$\Rightarrow \left. \frac{d\Delta_b}{d\delta_b} \right|_{d\delta_s = 0} > 0. \tag{5.30}$$

Total differentiation of (5.17) under using (5.30) gives

$$\left. \frac{dU_b}{d\delta_b} \right|_{d\delta_s = 0} = \frac{\partial U_b}{\partial \delta_b} + \frac{\partial U_b}{\partial \Delta_b} \left. \frac{d\Delta_b}{d\delta_b} \right|_{d\delta_s = 0},$$

$$\left. \frac{dU_s}{d\delta_b} \right|_{d\delta_s = 0} = \frac{\partial U_s}{\partial \delta_b} + \frac{\partial U_s}{\partial \Delta_b} \left. \frac{d\Delta_b}{d\delta_b} \right|_{d\delta_s = 0}. \tag{5.31}$$

Both total differentials are also ambiguous in sign: $\partial U_b / \partial \delta_b$ is negative but $\partial U_b / \partial \Delta_b \ d\Delta_b / d\delta_b$ is positive, $\partial U_s / \partial \delta_b$ is positive but $\partial U_s / \partial \Delta_b \ d\Delta_b / d\delta_b$ is negative. Because of the logarithmic terms (see (5.30)), an explicit treatment of (5.31) cannot resolve the ambiguity in signs.

Note that this section has implicitly assumed that both agents' optimal policies according to (5.12) and (5.13) are feasible. If feasibility is not given because both (5.20) and (5.21) are not met by the discount factors, the intersection of both agents' compliance constraints defines both agents' optimal policies. This case is depicted in Figure 5.7 which combines Figures 5.4 and 5.5.

Obviously, an increase (decrease) of one agent's discount factor increases (decreases) both agents' utilities because it allows for more (less) cooperation. Unless an increase lets both agents' optimal policies fall apart, more scope for cooperation benefits both the buyer and the seller in this case.

Some simulations can pronounce the salient results of strategic bargaining in the buyer–seller case. Tables 5.1 and 5.2 assume that $\alpha_b = \gamma_s = 1$ holds. They assume that δ_s is set at 0.7 and 0.9, respectively, and consider variations of the buyer's discount factor.

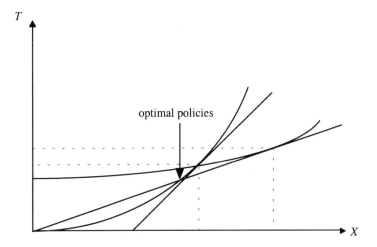

Figure 5.7   Coincidence of optimal policies

In both tables, p indicates the price per unit due to the strategic bargaining solution for self-enforcing contracts and p^* indicates the price per unit which would result from strategic bargaining for enforceable contracts. Because of the specification of parameters, p and p^* coincide if the strategic bargaining solution is to be found on the seller's compliance constraint (and, of course, in general if τ gives the strategic bargaining solution).

Since in both Tables 5.1 and 5.2 strategic bargaining solutions employ Δ_b or Δ_s in nearly all cases, these tables show that a strategic bargaining solution which requires that both agents agree upon the Pareto-optimal production level (which is one due to $\alpha_b = \gamma_s = 1$) is rather more the exception than the rule. For $\delta_s = 0.7$, only $\delta_b = 0.75$, for $\delta_s = 0.9$, only $\delta_b \in \{0.85, 0.9\}$ imply the Pareto-optimal production level. $\delta_s = 0.7$ and $\delta_b = 0.75$ show also that the impacts of both agents' discount factors on the agreement is quite different because $\delta_s = \delta_b = 0.7$ does not imply τ as the bargaining parameter but a strategic bargaining solution on the buyer's compliance constraint. In both tables, an increase in the buyer's discount factor starts increasing and continues decreasing his/her utility. The price p never exceeds the price p^*.

Table 5.1 Bargaining parameters and bargaining gains for $\delta_s = 0.7$ and different discount factors of the buyer (changed from 0.3 to 0.9) in the buyer–seller case for $\alpha_b = \gamma_s = 1$

δ_b	Bargaining parameter	U_b	U_s	p	p^*
0.3	$\Delta_b = 0.338704$	0.237093	0.044251	0.3	0.885729
0.35	$\Delta_b = 0.400822$	0.260523	0.059959	0.35	0.873205
0.4	$\Delta_b = 0.465168$	0.279101	0.077877	0.4	0.859904
0.45	$\Delta_b = 0.532154$	0.292685	0.097875	0.45	0.84562
0.5	$\Delta_b = 0.602318$	0.301159	0.119765	0.5	0.830126
0.55	$\Delta_b = 0.676371$	0.304367	0.143265	0.55	0.813164
0.6	$\Delta_b = 0.755264$	0.302106	0.167949	0.6	0.794424
0.65	$\Delta_b = 0.840306$	0.294107	0.193142	0.65	0.773528
0.7	$\Delta_b = 0.933333$	0.28	0.217778	0.7	0.75
0.75	$\tau = 0.723232$	0.276768	0.223232	0.723232	0.723232
0.8	$\Delta_s = 0.969395$	0.298161	0.20137	0.692425	0.692425
0.85	$\Delta_s = 0.919115$	0.315706	0.181023	0.656511	0.656511
0.9	$\Delta_s = 0.859625$	0.3318	0.158347	0.614018	0.614018

Table 5.2 Bargaining parameters and bargaining gains for $\delta_s = 0.9$ and different discount factors of the buyer (changed from 0.3 to 0.9) in the buyer–seller case for $\alpha_b = \gamma_s = 1$

δ_b	Bargaining parameter	U_b	U_s	p	p^*
0.3	$\Delta_b = 0.312576$	0.218803	0.044921	0.3	0.959766
0.35	$\Delta_b = 0.366724$	0.238371	0.06111	0.35	0.954397
0.4	$\Delta_b = 0.421747$	0.253048	0.079764	0.4	0.948436
0.45	$\Delta_b = 0.477851$	0.262818	0.100862	0.45	0.941717
0.5	$\Delta_b = 0.535317$	0.267658	0.124376	0.5	0.934027
0.55	$\Delta_b = 0.59454$	0.267543	0.150258	0.55	0.925085
0.6	$\Delta_b = 0.656092$	0.262437	0.178427	0.6	0.914506
0.65	$\Delta_b = 0.720827$	0.252289	0.208742	0.65	0.901742
0.7	$\Delta_b = 0.790084$	0.237025	0.240942	0.7	0.885982
0.75	$\Delta_b = 0.866083$	0.216521	0.274512	0.75	0.865968
0.8	$\Delta_b = 0.952794$	0.190559	0.308327	0.8	0.839636
0.85	$\tau = 0.803343$	0.196657	0.303343	0.803343	0.803343
0.9	$\tau = 0.75$	0.25	0.25	0.75	0.75

Tables 5.3 and 5.4 take the opposite position and fix δ_b at 0.7 and 0.9, respectively, and change the seller's discount factor. The picture is quite different compared to Tables 5.1 and 5.2: $\delta_b = 0.7$ does not allow the Pareto-optimal level to be realized but determines some strategic bargaining solutions as the intersection of both agents' compliance constraints. $\delta_b = 0.9$ implies strategic bargaining solutions which realize the Pareto-optimal level in three cases. The scope for these solutions, however, is obviously similarly restricted as in the first two tables.

Table 5.3 Bargaining parameters and bargaining gains for $\delta_b = 0.7$ and different discount factors of the seller (changed from 0.3 to 0.9) in the buyer–seller case for $\alpha_b = \gamma_s = 1$

δ_s	Bargaining parameter	U_b	U_s	p	p^*
0.3	$\Delta_s = 0.368563$	0.142165	0.158478	0.614271	0.614271
0.35	$\Delta_s = 0.438757$	0.163746	0.178757	0.626795	0.626795
0.4	$\Delta_s = 0.512077$	0.184298	0.196667	0.640096	0.640096
0.45	$\Delta_s = 0.588942$	0.20355	0.211966	0.65438	0.65438
0.5	$\Delta_s = 0.669874$	0.221143	0.224366	0.669874	0.669874
0.55	$\Delta_s = 0.75552$	0.236601	0.233513	0.686836	0.686836
0.6	intersection of both agents'	0.252	0.2352	0.7	0.705576
0.65	compliance constraints	0.273	0.22295	0.7	0.726472
0.7	$\Delta_b = 0.933333$	0.28	0.217778	0.7	0.75
0.75	$\Delta_b = 0.90117$	0.270351	0.224765	0.7	0.776768
0.8	$\Delta_b = 0.866793$	0.260038	0.23109	0.7	0.807575
0.85	$\Delta_b = 0.829886$	0.248966	0.236565	0.7	0.843489
0.9	$\Delta_b = 0.790084$	0.237025	0.240942	0.7	0.885982

Tables 5.3 and 5.4 show also that the price p never exceeds the price p^*. The asymmetry of the different impacts of discount factor can be demonstrated by comparing the bargaining gains in Table 5.1 or Table 5.2 with those in Table 5.3 or Table 5.4, respectively. Except for the cases of identical discount factors, the bargaining gains which the buyer (the seller) enjoys in Tables 5.1 and 5.2 exceed (fall short of) those he/she enjoys in Tables 5.3 and 5.4. This result demonstrates that changing the roles of the two agents does not produce a mirror image of the original situation as it was valid for many situations in Chapters 3 and 4. Significant differences arise in this setting because the buyer enjoys constant marginal benefits whereas the

seller faces increasing marginal costs. Obviously, the strategic bargaining solution also depends strongly on the shape of the assumed functions in addition to relative impatience and assumed parameters.

Table 5.4 Bargaining parameters and bargaining gains for $\delta_b = 0.9$ and different discount factors of the seller (changed from 0.3 to 0.9) in the buyer–seller case for $\alpha_b = \gamma_s = 1$

δ_s	Bargaining parameter	U_b	U_s	p	p^*
0.3	$\Delta_s = 0.324141$	0.149029	0.122578	0.540234	0.540234
0.35	$\Delta_s = 0.381922$	0.173544	0.135446	0.545603	0.545603
0.4	$\Delta_s = 0.441251$	0.197873	0.146027	0.551564	0.551564
0.45	$\Delta_s = 0.502455$	0.221943	0.154282	0.558283	0.558283
0.5	$\Delta_s = 0.565973$	0.245648	0.160163	0.565973	0.565973
0.55	$\Delta_s = 0.632407$	0.268826	0.163611	0.574915	0.574915
0.6	$\Delta_s = 0.702593$	0.291229	0.164546	0.585494	0.585494
0.65	$\Delta_s = 0.777735$	0.312449	0.16285	0.598258	0.598258
0.7	$\Delta_s = 0.859625$	0.3318	0.158347	0.614018	0.614018
0.75	$\Delta_s = 0.951048$	0.348053	0.150749	0.634032	0.634032
0.8	$\tau = 0.660364$	0.339636	0.160364	0.660364	0.660364
0.85	$\tau = 0.696657$	0.303343	0.196657	0.696657	0.696657
0.9	$\tau = 0.75$	0.25	0.25	0.75	0.75

5.4 SUMMARY

The discussion of the buyer–seller case benefited from the discussions in the previous chapters which facilitated developing the salient results of strategic bargaining in this setting:

- In the buyer–seller case, utilities and bargaining gains coincide because no action is taken by each agent in a one-shot Nash equilibrium: the buyer does not pay for nothing and the seller does not produce without receiving transfers.

  Both Chapters 3 and 4 employed a model which lets both agents introduce strictly positive effort levels in the one-shot game. In the buyer–seller case, efforts of each agent are zero for the purely non-cooperative equilibrium. Hence, international contracts and contracts which had to carry prohibitively high enforcement costs will only be agreed upon if repetition can prevent non-compliance. If potential exchanges are definitely limited to one period (or to a finite number of periods), no exchange will take place.

- Strategic bargaining may specify a self-enforcing contract which lies on a compliance constraint curve even if the Pareto-optimal level is feasible. Additionally, the buyer–seller case repeats some results of bargaining for a collective good because optimal policies do not specify the Pareto-optimal level, and the impact of discount factor changes is also ambiguous if the discount factor which defines the relevant range of the concession line is changed.

  These results have straightforward implications for international contracts. Strategic bargaining between a foreign investor and a country may lead to underinvestment if either the investor or the country has a comparably strong bargaining position. Underinvestment may result from the bargaining process although higher investment levels would not lead to non-compliance of an agent. But they may not match the strong bargaining position of either the investor or the country. When comparing different investments in different countries, an investor is likely to prefer to enter into a bargaining process with a partner with medium bargaining power because an institutionally weak partner implies high compliance costs.

- Different functional specifications of the buyer's benefits and the seller's production costs determine different conditions for the agents' discount factors in order to ensure feasibility.

  This result emphasizes that different functional specifications do not allow the behaviour of one partner to be given as the mirror image of the other partner. A producer who faces strictly increasing marginal costs imposes stricter restrictions on compliance than a producer who faces constant marginal costs.

6. Concluding Remarks

The model of strategic bargaining for self-enforcing contracts determined a unique solution for an important class of economic games which are subject to strategic behaviour. It has demonstrated that the relevant agreement for long-run cooperation may be determined by a simple bargaining procedure. It depends on the utility functions and the institutional toughness of both agents. In the preview of this book, several questions were asked for which this book has promised to give an answer. These answers may be summarized now:

- What role do contracts play in a non-cooperative environment in which the economic game does not guarantee compliance?

  Contracts play quite a similar role to that in the case of enforceable agreements. Self-enforcing contracts serve for strategy specification as do enforceable contracts. They support the agreement by their own means because they specify the actions to be taken in the case of non-compliance.

- What conditions should a contract fulfil to qualify as self-enforcing, that is, for an instrument which sustains long-run cooperation?

  It was shown that a self-enforcing contract should meet the conditions of weak renegotiation-proofness. If a contract is renegotiation proof, it is immune to renegotiations after defection has occurred. Additionally, it allows a return to cooperation after defection.

- Which self-enforcing contract will be chosen?

  The conditions that a self-enforcing contract should fulfil define a concession line which gives all Pareto-undominated feasible policies. Strategic bargaining determines a unique solution out of the concession line under fairly general conditions.

- How does the institutional setting influence the outcome?

  The impact of a discount factor change is ambiguous if the discount factor of the agent whose compliance constraint defines the part of the concession line under consideration is changed. On the one hand, the relative bargaining power is changed; on the other hand, the compliance constraint itself is changed. Both effects were shown to work in opposite directions. The section on policy implications will return to this case. In all other cases, only the bargaining effect which decreases the utility of the agent whose relative impatience is decreased, applies.

- Is a Pareto-optimal contract always chosen if possible?

     Since each agent aims at maximizing his/her bargaining gains, Pareto-optimal outcomes which enforceable contracts would specify are not necessarily agreed upon by a self-enforcing contract even if they are feasible.

- Do transfers always improve on the corresponding outcome without transfers?

     It was shown that every transfer policy must be self-enforcing as well. Then, transfers are likely to improve on the outcome without transfers when the benefits and costs differ substantially between both agents. They are not always introduced but the agreement may refrain from transfers even if they are allowed.

- Might a party prefer a self-enforcing contract even if enforceable contracts were possible?

     A party prefers a self-enforcing contract if the other party has a strong bargaining power. In this case, an enforceable contract would give the other party very low bargaining gains. A self-enforcing contract must be made credible by sufficient bargaining gains which prevent each party from breaking the contract. Hence, a party with little bargaining power may prefer the self-enforcing contract which guarantees a certain level of bargaining gains for reasons of compliance.

Chapters 3, 4 and 5 have demonstrated what strategic bargaining for a self-enforcing contract implies for concrete applications. A selective survey includes the results for implicit collusion in an industrial market, for the chances of free trade and for international contracts:

A duopolist's best policy does not leave the other duopolist on his/her non-cooperative profit level but increases his/her profits in order to make defection from implicit collusion unprofitable. Therefore, a duopolist will not prefer to join a partnership for implicit collusion with a very impatient partner because his/her costs for guaranteeing compliance of this partner are very high. Instead, he/she will seek cooperation with a more patient partner whose discount factor balances bargaining and compliance effect.

The impact of strategic bargaining on the chances of realizing Pareto-optimal levels was demonstrated for trade policies. The Pareto-optimal levels of trade policies require all trade barriers to be abolished because free trade maximizes the sum of both countries' welfare. Each country is interested in maximizing not world welfare but its individual welfare subject to compliance of the other country and potential delay costs in strategic bargaining. If one country had a comparably strong bargaining position, it would be able to seize a huge part of the gains from free trade of the other country if all trade barriers were to be abolished. This division of gains, however, makes non-compliance of the other country more likely the stronger is the bargaining power of its partner. It was shown on the basis of two numerical examples that free trade is unlikely to qualify for a strategic bargaining solution if bargaining powers between countries differ.

If transfers may play a role, it is not guaranteed that they will be employed in every case. Thus, one may not conclude that an agreement which specifies no transfers although transfers are allowed is an irrational outcome. Instead, it may mirror the result of strategic bargaining between two partners who deliberately do not employ transfers.

In the buyer–seller case, the effort of each agent is zero for the purely non-cooperative equilibrium. Hence, international contracts and contracts which had to carry prohibitively high enforcement costs will only be agreed upon if repetition may prevent non-compliance. If potential exchanges are definitely limited to one period (or a finite number of periods), no exchange will take place. In the case of infinite repetition, strategic bargaining between a foreign investor and a country may lead to underinvestment if either the investor or the country has a comparably strong bargaining position. Underinvestment may result from the bargaining process although higher investment levels would not lead to non-compliance of any agent. But they may not match the strong bargaining position of either the investor or the country. When comparing different investments in different countries, an investor is likely to prefer to enter into a bargaining process with a partner with medium bargaining power because an institutionally weak partner implies high compliance costs.

As the institutional setting was not discussed when deriving these conclusions, some policy conclusions may be drawn. These policy conclusions endogenize the part of the institutional setting which may be influenced by policy makers.

6.1 POLICY IMPLICATIONS

It should be clearly recognized that discussing policy implications depends crucially on the relationship between those agents who bargain for a self-enforcing contract and those agents who decide on the institutional setting. For example, two duopolists bargaining for implicit collusion do not determine the institutional setting which is set by the government and its authorities. Conversely, two countries bargaining for trade policies are also the agents who decide on the institutional setting, in particular when they can commit themselves to cooperation by restricting national sovereignty. Table 6.1 gives the distinction for the applications mentioned in the first sections of Chapters 3, 4 and 5.

Table 6.1 Role of agents bargaining for a self-enforcing contract and/or deciding on the institutional setting

Bargaining agents decide on the institutional setting, too:	Bargaining agents and agents deciding on the institutional setting differ:
Trade policy	Antitrust policy
International environmental policy	International project management
International project management	Contract law
	Civil law

International project management may concern either two countries (first column) or two firms in two different countries (second column). The term contract law mirrors all policy efforts to restrict illegal contracts. Specification of civil law determines the factual enforceability or the need for self-enforceability of contracts which are all theoretically enforceable (but may incur prohibitively high enforcement costs). If the agents bargaining for a self-enforcing contract and the agents deciding on the institutional setting differ, the relationship between both groups' policy intentions depends on the policy objectives. A government can be expected to aim at restricting collusion and illegal contracts. Conversely, a government may support international project management in order to increase welfare and employment in its country. Thus, the policy intentions of governments and those of an agent under consideration may be opposite or parallel.

Of course, policy implications do not suggest that an agent should increase or decrease his/her discount factor. Discount factor specification is exogenous. For example, a government representing a country in international trade negotiations shows a degree of impatience which depends on its own assessment of future risks and the pressure imposed by the domestic public to produce quick results. Policy implications address the institutional setting which was assumed to be given for strategic bargaining for self-enforcing contracts.

When the bargaining agents decide on the institutional setting, one could expect that they do their best to guarantee enforceability. If both contractors guaranteed enforceability, no contractor would be expected to refrain from making contracts enforceable. In some cases, however, it is surprising that contracts must be self-enforcing even though enforceability would raise no unsolvable problems. Trade policy, international environmental policy and some forms of international project management also give the agents who bargain for a contract the power to determine the institutional setting. For example, countries may deliberately restrict their sovereignty by joining a union of states which governs trade policy issues by a common law. But all the chapters have demonstrated that a relatively impatient agent is rewarded for his/her impatience by a high effort level of the other agent which makes his/her non-compliance non-profitable. He/she might expect lower bargaining gains if contracts are enforceable and if the other agent is relatively patient and could exploit his/her bargaining power without restrictions imposed by compliance constraints. Therefore, one agent may deliberately refuse enforceability in order to ensure a certain level of bargaining gains. This result may hold even if enforceable contracts may cover the infinite time horizon and may exploit intertemporal gains (see Section 2.6).

Chapter 2 demonstrated that antitrust policies cannot eliminate collusion unless collusive strategies undertaken by firms can be verified as anti-competitive. In the standard cases, restricting collusion increases the sum of consumer and producer rents. It can be successful if the risks of repetition are increased, for example by encouraging a third party to enter. If future collusion is at risk, future punishment is as well and non-compliant behaviour is more attractive. If a third party enters the market, however, perfectly non-cooperative behaviour is not guaranteed if collusion is possible among all firms or among the members of a coalition.

With respect to contract law and civil law, two conclusions can be drawn. First, prohibition of contracts cannot be completely successful, and deficiencies in enforcement by civil law do not result in purely non-cooperative outcomes unless repetition can be excluded. Both results are due to some scope of self-enforcing cooperation which is exploited by strategic bargaining for the respective contract. Second, abolishing prohibition and lifting the frontier between enforceability and self-enforceability may change not only total gains but also distributional patterns. For example, civil law may decrease transaction costs for enforcing contracts. In this case, some contracts – which formerly had to be self-enforcing – now become enforceable because bringing non-compliant behaviour to court is now profitable. Changed transaction costs alter individual bargaining gains and are likely to decrease the relatively impatient agent's gains because the other agent no longer has to sacrifice bargaining gains for guaranteeing compliance.

Chapters 3 and 4 dealt with basically the same problem but differed with respect to the role that transfers were allowed to play. Chapter 3 assumed that the institutional setting aims at restricting cooperation whereas Chapter 4 assumed that cooperation is appreciated by policy makers. The results of Chapter 3 relied on the assumption that prohibition of transfers is perfect. In general, prohibition of transfers raises control costs which must be balanced by control benefits. Suppose that control benefits are the higher the lower are the sum of both agents' bargaining gains (for the case of a duopoly, restricting collusion benefits consumers and increases the sum of consumers' and producers' rents). For the optimal policies, it has been demonstrated that both agents' individual bargaining gains may be decreased or increased by allowing transfers if the agents' costs and benefits are rather similar. It has also been shown that both agents' individual bargaining gains are likely to be increased if benefits and costs differ substantially. Thus, control of transfers should be concentrated mainly on potential cooperation between agents who carry significantly different costs and enjoy significantly different benefits. Conversely, control of transfers may only negligibly improve on the control objective if agents are identical because the impact on the performance is small compared to cooperation with transfers in these cases. This result obviously holds very effectively for the buyer–seller case because a complete control of transfers chokes off any cooperation between buyer and seller.

6.2 AN AGENDA FOR FUTURE RESEARCH

This book has explored the scope for cooperation in a framework which has faced a twofold restriction: first, all models have assumed that neither the environment nor the utility functions (nor the discount factors) change in the course of time; second, the analysis has been restricted to two agents. It is evident that future research should weaken these assumptions. The specific assumptions made here were due to the substantial degree of complexity and ambiguity which a more general approach is likely to face. Therefore, future research should deal with a specific generalization instead of trying several generalizations simultaneously. The following generalizations may deserve further investigation:

- When giving up the stationarity assumption that repeated games employ, time-consistent bargaining behaviour will not imply identical outcomes for every period. Instead, the intertemporal path should not be subject to revision. This condition is satisfied if all environments which are reached by coordinated behaviour guarantee that agreed-upon future actions are the result of strategic bargaining as if bargaining has started then. In this case, no agent will prefer to submit an alternative proposal. This condition may be expected to restrict the set of attainable outcomes substantially. It could be checked for different stock–flow relationships, first, by assuming non-binding compliance constraints. For example, one could discuss international environmental problems by assuming that the actions of two countries determine a stock which itself enters the social welfare function of each country. This exercise resembles the time-consistency problem of optimal policies, for example optimal monetary policies (the shaping papers of this strand of literature are Barro and Gordon, 1983 and Kydland and Prescott, 1977). For this exercise, it is necessary to determine the non-cooperative path in order to be able to determine the bargaining gains. Note that a selection problem arises because several solution concepts for dynamic games exist which imply different paths. Hence, even if every stage game is of the prisoners' dilemma type, the non-cooperative path may be indeterminate without further assumptions of dynamic behaviour. If credibility considerations require that strategies of a non-cooperative equilibrium should be Markov perfect, Markov–perfect equilibria may not exist even though subgame-perfect equilibria do (see Hellwig and Leininger, 1988).
- In another approach, the definition of compliance constraints for dynamic modelling could be discussed. As the situation is changed, the compliance constraints differ between periods. It could be interesting to explore the behaviour of compliance constraints in time, in particular whether they are likely to bind more in current or future periods. Since Section 2.2 has demonstrated that the role of the discount factor is ambiguous in dynamic games, one should not expect clear-cut results.
- A third approach could address the problem of bargaining for a self-enforcing contract among three agents in a stationary world. Discussing cooperation among three agents involves dealing with two problems. First, the role of coalitions covering only two agents must be explored. The coalition plays a role because two out of these three agents may join a partnership without the third agent or agree upon joint defection from

overall cooperation. Second, a bargaining model must be employed which leads to a unique solution in this setting.

Future research could address specific applications, for example by studying international environmental problems using a dynamic model. Particular applications allow the dynamics of the model to be restricted and specified. For example, Section 2.2 discussed a resource extraction model which employed a simple stock–flow relationship. Another dynamic extension could address multilateral trade agreements. In addition to considering more than two countries, it could be interesting to discuss cooperation in a dynamic model which endogenizes capital accumulation: if the degree of investment depends on the current wealth of a country which itself depends on the degree of self-enforcing trade liberalization, all countries enter a dynamic game. It might be interesting to explore whether the chances for trade liberalization increase or decrease in time.

This book has set the stage for all research proposals mentioned in this agenda for future research by studying the basic model. If the basic model has encouraged some readers to employ this approach for studying further applications or to work on generalizations, then the book has achieved its goal.

Appendix: Rebargaining-proofness

Section 2.4 discussed strong renegotiation-proofness and rebargaining-proofness. This appendix will present the exact definitions of both strong renegotiation-proofness and rebargaining-proofness for the models under consideration. Strong renegotiation-proofness requires that it should not pay for both agents to switch to another weakly renegotiation-proof outcome instead of following the punishment plan of the old agreement. Let $[\bar{s}_i, \bar{s}_j]$ denote a strategy combination which qualifies for a self-enforcing contract. Hence, $[\bar{s}_i, \bar{s}_j]$ is weakly renegotiation proof. Let $[\tilde{\bar{s}}_i, \tilde{\bar{s}}_j]$ denote any other strategy combination. $[\bar{s}_i, \bar{s}_j]$ is strongly renegotiation proof if (A.1) holds:

$$\forall k \neq -k \in \{i, j\}, \quad \forall \left[\tilde{\bar{s}}_i, \tilde{\bar{s}}_j\right]:$$

$$\frac{1}{1-\delta_k} U_k\left[\tilde{\bar{s}}_i, \tilde{\bar{s}}_j\right] - \frac{1-\delta_k^n}{1-\delta_k} U_k\left[\bar{s}_k, s'_{-k}\left(\bar{s}_k\right)\right] - \frac{\delta_k^n}{1-\delta_k} U_k\left[\bar{s}_k, \bar{s}_{-k}\right] \geq 0,$$

$$\frac{1}{1-\delta_{-k}} U_{-k}\left[\tilde{\bar{s}}_i, \tilde{\bar{s}}_j\right] - \frac{1-\delta_{-k}^n}{1-\delta_{-k}} U_{-k}\left[\bar{s}_k, s'_{-k}\left(\bar{s}_k\right)\right] - \frac{\delta_{-k}^n}{1-\delta_{-k}} U_{-k}\left[\bar{s}_k, \bar{s}_{-k}\right] \geq 0$$

$$\Rightarrow \left[\tilde{\bar{s}}_i, \tilde{\bar{s}}_j\right] \text{ does not belong to the set of efficient agreements.} \qquad (A.1)$$

Condition (A.1) requires that every agreement which successfully replaces the old punishment plan (see the first terms in the second and third line of (A.1)) because it gives both the agent to be punished (see the second line) and the other agent (see the third line) higher benefits compared to the old punishment plan should not belong to the set of efficient agreements. The implications of strong renegotiation-proofness can be exemplified by Table A.1, which reproduces Table 2.1 and adds another option for each agent.

Assumption (A.2) defines the range of parameters used in Table A.1.

$$a, b, d, e, f, g > 0, \quad c, h < 0, \quad a < b, \quad a > d, \quad a < e, \quad a > f, \quad b > e, \quad f < g. \qquad (A.2)$$

Table A.1 Payoffs from cooperation and non-cooperation for three options
       for agent i and agent j

Agent *i* chooses rows Agent *j* chooses columns	I	II	III
I	(a,a)	(f,e)	(c,b)
II	(e,f)	(d,d)	(h,g)
III	(b,c)	(g,h)	(0,0)

Assumption (A.2) ensures that the unique one-shot Nash equilibrium is (III,III). The Pareto-optimal outcomes are given by the payoff combinations (a,a), (e,f), (f,e). (d,d) improves on the one-shot outcome but is Pareto dominated by (a,a). Suppose that the punishment plan restricts punishment to one period. The conditions for weak renegotiation-proofness of the Pareto-optimal outcomes and the Pareto-inferior outcome (d,d) are given by:

$$\delta \geq \max\left\{\frac{b-a}{a-c}, \frac{-c}{a-c}\right\}, \quad \delta \geq \max\left\{\frac{b-e}{e-h}, \frac{-h}{e-h}\right\}$$

$$\delta \geq \max\left\{\frac{g-f}{f-c}, \frac{-c}{f-c}\right\}, \quad \delta \geq \max\left\{\frac{g-d}{d-h}, \frac{-h}{d-h}\right\}. \qquad (A.3)$$

Condition (A.3) can be derived using (2.17) and (2.19). The first condition of (A.3) gives the condition for weak renegotiation-proofness of (a,a), the last condition gives the respective condition for (d,d). Both agreements are perfectly symmetric because both agents' defection option is the same. Symmetry does not hold for (e,f) and (f,e) because both agents' defection options fall apart. The second and the third conditions give the conditions for (e,f) and (f,e): the second condition ensures the cooperative behaviour of agent *j* (*i*) for (e,f) [(f,e)], the third condition ensures the cooperative behaviour of agent *i* (*j*) for (e,f) [(f,e)].

Strong renegotiation-proofness considers the case that an agent has deviated from the agreed-upon cooperation and makes a proposal which substitutes for punishment and a resumption of the old agreement. Suppose that the deviating agent *i* always proposes (e,f) if he/she has deviated from

(a,a), (b,b) or (f,e). This policy is arbitrarily chosen and other policies might be taken into consideration as well but a strongly renegotiation-proof agreement should at least be immune to the respective new agreement. Neglecting deviation from (e,f) does not make these considerations incomplete: if (e,f) is agreed upon, the policy of the potentially deviating agent j with respect to deviation and a new proposal are the same as agent i's policy concerning (f,e). Since renegotiation-proofness should hold for both agents, concentrating on (a,a), (b,b) and (f,e) determines some necessary conditions for strong renegotiation-proofness.

An alternative agreement which should successfully replace punishment and resume the old agreement must qualify for Pareto superiority: agent i must be better off under the new agreement instead of investing one period in a resumption of the old agreement, agent j must be better off under the new agreement instead of enjoying 'allowed' defection in one period and a resumption of the old agreement. For the alternative proposals as indicated above, agent i makes these proposals if (A.4) holds:

$$c + \frac{\delta a}{1-\delta} \leq \frac{e}{1-\delta} \Leftrightarrow \delta \leq \frac{e-c}{a-c},$$

$$c + \frac{\delta f}{1-\delta} \leq \frac{e}{1-\delta} \Leftrightarrow \delta \leq \frac{e-c}{f-c},$$

$$h + \frac{\delta d}{1-\delta} \leq \frac{e}{1-\delta} \Leftrightarrow \delta \leq \frac{e-h}{d-h}. \tag{A.4}$$

The first terms on the LHS give the punishment to be borne by agent i; the second terms give the discounted value of a resumption of cooperation according to the old agreement. The RHS gives the value of the new agreement proposed to replace the old agreement.

Agent j accepts these proposals if (A.5) holds:

$$b + \frac{\delta a}{1-\delta} \leq \frac{f}{1-\delta} \Leftrightarrow \delta \geq \frac{f-b}{a-b},$$

$$b + \frac{\delta e}{1-\delta} \leq \frac{f}{1-\delta} \Leftrightarrow \delta \geq \frac{f-b}{e-b},$$

$$g + \frac{\delta d}{1-\delta} \leq \frac{f}{1-\delta} \Leftrightarrow \delta \geq \frac{f-g}{d-g}. \tag{A.5}$$

The first terms on the LHS give the benefits of allowed defection enjoyed by agent j; the second terms give the discounted value of a resumption of cooperation according to the old agreement. The RHS gives the value of the new agreement proposed to replace the old agreement. If a certain agreement is not weakly renegotiation proof, that is, one of the lines of (A.3) does not hold, it cannot be a candidate for a strongly renegotiation-proof agreement. If (A.4) and (A.5) are fulfilled for all agreements which are weakly renegotiation proof according to (A.3), none of them is strongly renegotiation proof. In this case, at least one agent is able to make a proposal which will successfully replace his punishment. Strong renegotiation-proofness would imply that no agreement is credible because it is either not supported by the punishment plan or its punishment plan is dominated by another agreement. The fatal implication would be that no agreement which improves on the one-shot Nash equilibrium is credible unless a credible commitment to punish every time is possible.

The following example demonstrates the implication of strong renegotiation-proofness. Table A.2 specifies the parameters introduced by Table A.1:

Table A.2 Example for Table A.1

Agent *i* chooses rows Agent *j* chooses columns	I	II	III
I	(5,5)	(6,4)	(–4,10)
II	(4,6)	(2,2)	(–4,10)
III	(10,–4)	(10,–4)	(0,0)

Suppose that the discount factor δ of both agents is equal to 4/5. Equation (A.3) determines the conditions for weak renegotiation-proofness for every attainable agreement which improves on (0,0). Inserting the parameters of Table A.2 into (A.3) gives (A.3') which determines the minimum discount factor which supports weak renegotiation-proofness for the respective agreement:

$$\delta \geq \frac{5}{9} \text{ to support } (a,a),$$

$$\delta \geq \frac{3}{4} \text{ to support } (e,f) \text{ and } (f,e),$$

$$\delta \geq \frac{4}{3} \text{ to support } (d,d). \tag{A.3'}$$

The second line of (A.3') collects the second and the third line of (A.3) which fall apart because of the asymmetry of non-compliance for (e,f) and (f,e). Obviously, no reasonable discount factor can support (d,d) which can therefore be excluded from the set of feasible agreements. If agent i has deviated either from (a,a) or (f,e), respectively, he/she is better off by proposing (e,f) if (A.4') holds:

$$\delta \leq \frac{8}{9} \text{ to replace } (a,a) \text{ by } (e,f),$$

$$\delta \leq \frac{4}{5} \text{ to replace } (f,e) \text{ by } (e,f). \tag{A.4'}$$

Agent j will accept this proposal instead of insisting on punishment if the new agreement (e,f) gives him/her a higher utility than punishing i and returning to (a,a) or (f,e), respectively:

$$\delta \geq \frac{4}{5} \text{ to replace } (a,a) \text{ by } (e,f),$$

$$\delta \geq \frac{2}{3} \text{ to replace } (f,e) \text{ by } (e,f). \tag{A.5'}$$

Conditions (A.4') and (A.5') are exactly fulfilled for $\delta = 4/5$ because δ is low enough to make the lower cooperative utility of the new agreement attractive for agent i and high enough to make agent j refrain from enjoying punishment. (a,a) is not strongly renegotiation proof because the deviating agent gains by switching to (e,f) whereas the other agent does not lose; (e,f) and (f,e) are not strongly renegotiation proof because the deviating agent does not lose by switching to (e,f) whereas the other agent is made better off.

Note that strong renegotiation-proofness considers only the incentive to propose a new agreement after defection has occurred. But even taking the *ex ante* incentives into account cannot save weakly renegotiation-proof contracts. Agent i can be expected to choose defection deliberately if the benefits of defection and the benefits arising from the new agreement exceed the benefits of keeping the old agreement:

$$\left. \begin{array}{l} b + \dfrac{\delta e}{1-\delta} > \dfrac{a}{1-\delta} \\[2mm] b + \dfrac{\delta e}{1-\delta} > \dfrac{f}{1-\delta} \\[2mm] g + \dfrac{\delta e}{1-\delta} > \dfrac{d}{1-\delta} \end{array} \right\} \Rightarrow \delta < \dfrac{a-b}{e-b} = \dfrac{5}{6}. \qquad (A.6)$$

The first terms on the LHS give the benefits from defection, the second terms give the discounted sum of benefits from the new agreement. The RHS indicates the discounted sum of benefits from the old agreement. Because of (A.2), the first condition dominates the other two conditions. Condition (A.6) is completely in line with $\delta = 4/5$ and the parameters of Table A.2. Thus, every agent has an incentive to deviate from every weakly renegotiation-proof agreement since he/she anticipates that punishment will not occur but an alternative agreement will be chosen unanimously. Condition (A.6) ensures that defection and switching to the alternative agreement benefits an agent more than keeping the original contract.

The example demonstrates that the refinement of strong renegotiation-proofness may severely restrict the scope of feasible cooperation. Alternative approaches have been developed which use other concepts of renegotiation-proofness and which allow more cooperation sustained by repetition. For example, Asheim (1991) criticized the fact that the demonstrated concept uses weakly renegotiation–proof agreements which are themselves not strongly renegotiation-proof to destroy a weakly renegotiation-proof agreement. Other approaches have been developed which introduce other concepts, for example social norms (for alternative approaches, see Asheim, 1991; Begin and MacLeod, 1993; Ray, 1994).

Introducing a social norm, however, conflicts with the definition of self-enforcing contracts in a non-cooperative environment which rules out the influence of any third party and of any exogenous rules. If self-enforcing contracts must rely on their own terms, there should be no restriction to submitting new proposals. If it is accepted that punishment can be questioned, weakly renegotiation-proof agreements destroyed by weakly renegotiation-proof agreements which are themselves not strongly renegotiation proof reflect consistent, not inconsistent behaviour. The concept of strong renegotiation-proofness requires only that agents should accept a weakly renegotiation-proof agreement when they have done so in the past. From the viewpoint of intertemporal consistency, accepting an agreement now but rejecting an agreement which satisfies the same conditions after a certain event (that is, defection) has occurred is more likely to reflect inconsistent behaviour.

However, strong renegotiation-proofness assumes implicitly that all agreements have the same chance of being selected. This is not the case if an agreement should be bargaining based. An agreement is bargaining based if it results from deliberate bargaining strategies and if bargaining for a certain division of bargaining gains leads to identical results for identical problems. In this case, there is no ambiguity about which outcome will be selected. Chapter 2 has introduced a strategic bargaining model which guarantees a bargaining-based agreement.

REBARGAINING-PROOFNESS

A bargaining-based, self-enforcing contract is called rebargaining proof of degree n if the rebargaining-based agreement which may substitute for punishment in the n punishment phases does not violate the condition of *ex ante* compliance. The rebargaining-based agreement is the result of the same bargaining procedure as the original agreement except that the utilities which the punishment plan produces serve as the utility reservation levels. Without going into details immediately, suppose that rebargaining leads to $[s_i^{**}, s_j^{**}]$ which fulfils the conditions of weak renegotiation-proofness according to (2.17) and (2.19). $[s_i^{**}, s_j^{**}]$ can only be a successful candidate for rebargaining if

$$U_{-k}\left[s_i^{**}, s_j^{**} \right] \geq U_{-k}\left[s'_{-k}\left(s_k^* \right), s_k^* \right] \tag{A.7}$$

holds. If (A.7) is violated, agent $-k$ is better off by insisting on punishment since any alternative cannot give him/her higher payoffs. If (A.7) is fulfilled, $[s_i^{**}, s_j^{**}]$ is rebargaining proof of degree n if (A.8) holds.

$$\frac{1-\delta_k^{n+1}}{1-\delta_k} U_k\left[s_k^*, s_{-k}^* \right] - U_k\left[s'_k\left(s_{-k}^* \right), s_{-k}^* \right] + \delta_k \frac{1-\delta_k^n}{1-\delta_k} U_k\left[s_k^{**}, s_{-k}^{**} \right] \geq 0. \tag{A.8}$$

Condition (A.8) reproduces (2.17) except that the punishment payoff is replaced by the payoff due to $[s_i^{**}, s_j^{**}]$. Condition (A.8) specifies that agent k has *ex ante* no incentive to break the contract and to rebargain which gives him/her $[s_i^{**}, s_j^{**}]$ during the punishment phase. Since $U_k[s_k^{**}, s_{-k}^{**}] > U_k[s_k^*, s'_{-k}(s_k^*)]$, (A.8) is more demanding than (2.17). Since the deviating agent is always weakly better off by rebargaining, *ex post* compliance is guaranteed in every case by (2.19).

One may now discuss in which cases rebargaining takes place and which result is to be expected if it does take place. Consider Figure A.1, which depicts the case of no rebargaining.

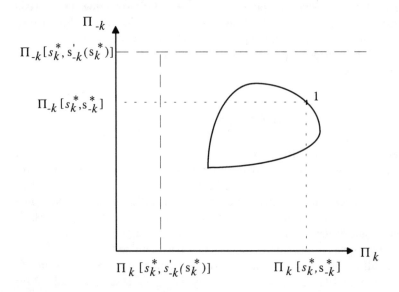

Figure A.1   No rebargaining

In Figure A.1, 1 depicts the strategic bargaining solution according to (2.52). The broken horizontal line gives the payoff which agent $-k$ enjoys during the punishment periods when punishment is carried out according to the second paragraph of the self-enforcing contract. The broken vertical line gives the respective payoff for agent k. Figure A.1 shows a case in which (A.7) cannot be satisfied because no feasible agreement exists which gives agent $-k$ his/her punishment payoff. In this case, rebargaining will not be successful.

Figure A.2 depicts the other case where the punishment benefits of agent $-k$ are dominated by other feasible agreements which belong to the set of efficient agreements. In this case, rebargaining will take place but is subject to the constraint that agent $-k$ is at least given his/her punishment payoff. In fact, this constraint binds. Since the general definition of rebargaining-proofness has specified that rebargaining employs the same bargaining procedure, unrestricted bargaining would give 1 again (recall that the delay costs are based on total payoffs). Hence, agent $-k$ accepts point 2 as the rebargaining solution and agent k enjoys $U_k[s_i^{**}, s_j^{**}]$ during the punishment periods. If point 2 satisfies (A.8), the rebargaining solution is rebargaining proof.

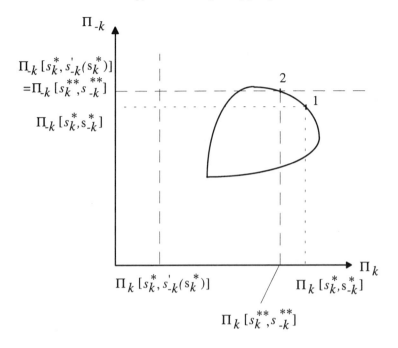

Figure A.2   Rebargaining and rebargaining solution

Rebargaining-proofness acknowledges that the rebargaining gains take into account the punishment originally specified by the second paragraph of the self-enforcing contract. Condition (A.7) specifies that rebargaining does not change the punishment plan if the agent who may enjoy *n* free rides cannot be made better off by any weakly renegotiation-proof contract. This situation is given by the example in Table A.2. All alternative agreements for one period make agent *j* worse off compared to punishing. (*a,a*), (*e,f*), and (*f,e*) are therefore rebargaining proof because there is no better world for this agent than to enjoy *b* or *g*, respectively. But even if rebargaining changes punishment, it is harmless unless an agent anticipating rebargaining is made better off by defection.

Table A.1 has assumed a specific punishment plan which was restricted to one period. This contract then would be rebargaining proof to degree 1. The definition of rebargaining-proofness emphasizes its degree because a higher degree is associated with longer punishment phases. Degree 1 should indicate that the agreement relies on only one punishment period. In general, the length of the punishment period is not given *ex ante* but determined to maximize the scope for cooperation. As all functions and parameters are common knowledge, the determination of *n* is straightforward. However, it might also be found that the number of punishment periods should be limited. The argument for an upper bound for n is that every punishment allows only a rebargaining, but not a bargaining of the basic issue. From this perspective, the credibility of rebargaining-proof contracts might be found to correspond with the number of punishment periods because a low number implies a credibly short period in which only rebargaining is possible. As the models in the previous chapters have determined $n = 1$, this aspect does not play any role there. However, if such a restriction is found necessary in general, one has to determine the binding constraint for $n = 1$ which is either (2.17) or (2.19).

The discussion of strong renegotiation-proofness and rebargaining-proofness should not lead to the conclusion that every bargaining-based self-enforcing contract is rebargaining proof. Rebargaining-proofness is not a redundant condition but may restrict the scope of cooperation as well. Suppose, for example, that the original agreement $[s_i^*, s_j^*]$ makes the *ex ante* compliance constraint equal to zero (as is the case for orthogonal reaction functions with $n = 1$ for which all compliance constraints coincide). If (A.7) is satisfied, $[s_i^*, s_j^*]$ is not rebargaining proof because (2.17) set equal to zero ensures that (A.8) is violated (because $U_k[s_k^{**}, s_{-k}^{**}] > U_k[s_k^*, s'_{-k}(s_k^*)]$). From this consideration, it is obvious that the chances for rebargaining-proofness are increased if the compliance constraints are overfulfilled. However, one cannot conclude on a general level that the chances for realizing a Pareto-optimal solution are increased. If (A.7) cannot be satisfied for the whole set of efficient agreements, every conceivable solution is rebargaining proof. Note that (A.7) does not guarantee strong renegotiation-proofness, as the example has demonstrated.

If the selected self-enforcing contract is not rebargaining proof, then there are three escape routes which can be taken by both agents. First, rebargaining may restrict the scope of feasible agreements in addition to weak renegotiation-proofness and may hence exclude several outcomes. Then, two additional constraints would have to be introduced. Second, rebargaining-proofness may be restored if both agents' punishment phases are differentiated. In this case, increasing n for one agent may make *ex ante* compliance from the viewpoint of rebargaining profitable without making the selected self-enforcing contract infeasible. Third, one could take the standpoint that punishment does not allow a rebargaining because punishment is assumed to conflict with communication.

References

Abreu, D. (1988), 'On the theory of infinitely repeated games with discounting', *Econometrica*, **56**, 383–96.

Abreu, D., Dutta, P.K. and Smith, L. (1994), 'The Folk Theorem for repeated games: a NEU condition', *Econometrica*, **62**, 939–48.

Abreu, D. and Pearce, D. (1991), 'A perspective on renegotiation in repeated games', in Selten, R. (ed.), *Game Equilibrium Models II: Methods, Morals, and Markets*, Berlin: Springer, pp. 44–55.

Abreu, D., Pearce, D. and Stacchetti, E. (1989), 'Renegotiation and symmetry in repeated games', Cowles Foundation Discussion Paper No. 920, Yale University.

Admati, A.R. and Perry, M. (1987), 'Strategic delay in bargaining', *Review of Economic Studies*, **54**, 345–64.

Arrow, K.J. and Debreu, G. (1954), 'Existence of an equilibrium for a competitive economy', *Econometrica*, **22**, 265–90.

Asheim, G.B. (1991), 'Extending renegotiation-proofness to infinite horizon games', *Games and Economic Behavior*, **3**, 278–94.

Atkeson, A. (1991), 'International lending with moral hazard and risk of repudiation', *Econometrica*, **59**, 1069–89.

Ausubel, L.M. and Deneckere, R.J. (1992), 'Bargaining and the right to remain silent', *Econometrica*, **60**, 597–625.

Barrett, S. (1991), 'The paradox of international environmental agreements', mimeo, London Business School.

Barrett, S. (1992), 'International environmental agreements as games', in Pethig, R. (ed.), *Conflicts and Cooperation in Managing Environmental Resources*, Berlin: Springer, pp. 11–36.

Barro, R.J. and Gordon, D.B. (1983), 'Rules, discretion and reputation in a model of monetary policy', *Journal of Monetary Economics*, **12**, 101–21.

Bauer, A. (1992), 'International cooperation over environmental goods', Münchener wirtschaftswissenschaftliche Beiträge 92.17, München.

Begin, J. and MacLeod, W.B. (1993), 'Efficiency and renegotiation in repeated games', *Journal of Economic Theory*, **61**, 42–73.

Benoit, J.-P. and Krishna, V. (1985), 'Finitely repeated games', *Econometrica*, **53**, 890–904.

Bernheim, B. and Dasgupta, A. (1995), 'Repeated games with asymptotically finite horizons', *Journal of Economic Theory*, **67**, 129–52.

Bernheim, B., Peleg, B. and Whinston, M. (1987), 'Coalition-proof Nash equilibria. I. Concepts', *Journal of Economic Theory*, **42**, 1–12.

Bernheim, B.D. and Ray, D. (1989), 'Collective dynamic consistency in repeated games', *Games and Economic Behavior*, **1**, 295–326.

Bester, H. (1989a), 'Non-cooperative bargaining and imperfect competition: a survey', *Zeitschrift für Wirtschafts- und Sozialwissenschaften*, **109**, 265–86.

Bester, H. (1989b), 'Noncooperative bargaining and spatial competititon', *Econometrica*, **57**, 97–113.

Bikhchandani, S. (1992), 'A bargaining model with incomplete information', *Review of Economic Studies*, **59**, 187–203.

Binmore, K. (1985), 'Bargaining and coalitions', in Roth (1985), pp. 269–304.

Binmore, K. (1987), 'Perfect equilibria in bargaining models', in Binmore and Dasgupta (1987b), pp. 77–105.

Binmore, K. and Dasgupta, P. (1987a), 'Nash bargaining theory: an introduction', in Binmore and Dasgupta (1987b), pp. 1–26.

Binmore, K. and Dasgupta, P. (eds) (1987b), *The Economics of Bargaining*, Oxford: Basil Blackwell.

Binmore, K., Osborne, M.J. and Rubinstein, A. (1992), 'Noncooperative models of bargaining', in Aumann, R.J. and Hart, S. (eds), *Handbook of Game Theory with Economic Applications*, Vol. 1, Amsterdam: North-Holland, pp. 179–225.

Binmore, K., Rubinstein, A. and Wolinsky, A. (1986), 'The Nash bargaining solution in economic modelling', *Rand Journal of Economics*, **17**, 176–88.

Böhm-Bawerk, E. von (1912), *Positive Theorie des Kapitals*, English translation: *Capital and Interest*, South Holland, Illinois: Libertarian Press, 1959.

Botteon, M. and Carraro, C., (1995), 'Burden-sharing and coalition stability in environmental negotiations with asymmetric countries', Nota di Lavoro 95.78, Fondazione ENI, Milan.

Brander, J.A. and Spencer, B. (1985), 'Export subsidies and international market share rivalry', *Journal of International Economics*, **18**, 83–100.

Bulow, J., Geanakoplos, J. and Klemperer, P. (1985), 'Multimarket Oligopoly: Strategic Substitutes and Complements', *Journal of Political Economy*, **93**, 488–511.

Bulow, J. and Rogoff, K. (1989), 'A constant recontracting model of sovereign debt', *Journal of Political Economy*, **79**, 155–78.

Carmichael, H.L. (1989), 'Self-enforcing contracts, shirking, and life cycle incentives', *Journal of Economic Perspectives*, **3**, 65–83.

Carraro, C. and Siniscalco, D. (1993), 'Strategies for the international protection of the environment', *Journal of Public Economics*, **52**, 309–28.

Cesar, H.S.J. (1994), *Control and Game Models of the Greenhouse Effect. Economic Essays on the Comedy and Tragedy of the Commons*, Berlin: Springer.

Chae, S. (1993), 'The *n*-person Nash bargaining solution with time preference', *Economics Letters*, **41**, 21–4.

Cramton, P.C. (1991), 'Dynamic bargaining with transaction costs', *Management Science*, **37**, 1221–33.

Cramton, P.C. (1992), 'Strategic delay in bargaining with two-sided uncertainty', *Review of Economic Studies*, **59**, 205–25.

Dewatripont, M. and Maskin, E. (1995), 'Contractual contingencies and renegotiation', *Rand Journal of Economics*, **26**, 704–19.

Dutta, P.K. (1995), 'Collusion, discounting and dynamic games', *Journal of Economic Theory*, **66**, 289–306.

Evans, R. and Maskin, E. (1989), 'Efficient renegotiation-proof equilibria in repeated games', *Games and Economic Behavior*, **1**, 361–9.

Fankhauser, S. (1995), *Valuing Climate Change: The Economics of the Greenhouse*, London: Earthscan Publications.

Farrell, J. and Gibbons, R. (1989), 'Cheap talk can matter in bargaining', *Journal of Economic Theory*, **48**, 221–37.

Farrell, J. and Maskin, E. (1989), 'Renegotiation in repeated games', *Games and Economic Behavior*, **1**, 327–60.

Fishburn, P. and Rubinstein, A. (1982), 'Time Preferences', *International Economic Review*, **23**, 677–94.

Friedman, J.W. (1971), 'A non-cooperative equilibrium for supergames', *Review of Economic Studies*, **38**, 1–12.

Fudenberg, D., Levine, D. and Maskin, E. (1994), 'The Folk Theorem with imperfect public information', *Econometrica*, **62**, 997–1039.

Fudenberg, D. and Maskin, E. (1986), 'The Folk Theorem in repeated games with discounting or with incomplete information', *Econometrica*, **54**, 533–54.

Fudenberg, D. and Tirole, J. (1991), *Game Theory*, Cambridge, Mass.: MIT Press.

Gandolfo, G. (1986), *International Economics*, Berlin: Springer.

Green, J.R. and Laffont, J.-J. (1992), 'Renegotiation and the form of efficient contracts', *Annales d'Economie et de Statistique*, **25/26**, 123–50.

Grossman, G.M. and Helpman, E. (1994), 'Protection for sale', *American Economic Review*, **84**, 833–50.

Grossman, G.M. and Helpman, E. (1995), 'Trade wars and trade talks', *Journal of Political Economy*, **103**, 675–708.

Gul, F. and Sonnenschein, H. (1988), 'On delay in bargaining with one-sided uncertainty', *Econometrica*, **56**, 601–11.

Güth, W. and Kalkofen, B. (1989), *Unique Solutions for Strategic Games. Equilibrium Selection Based on Resistance Avoidance*, Berlin: Springer.

Güth, W., Schmittberger, R. and Schwarze, B. (1982), 'An experimental analysis of ultimate bargaining', *Journal of Economic Behaviour and Organization*, **3**, 367–88.

Güth, W. and Tietz, R. (1990), 'Ultimate bargaining behaviour – a survey and comparison of experimental results', *Journal of Economic Psychology*, **11**, 417–49.

Harsanyi, H. and Selten, R. (1988), *A General Theory of Equilibrium Selection in Games*, Cambridge, Mass.: MIT Press.

Hart, O. and Moore, J. (1988), 'Incomplete contracts and renegotiation', *Econometrica*, **56**, 755–85.

Heal, G. (1992), 'International negotiations on emission control', *Structural Change and Economic Dynamics*, **3**, 223–40.

Hellwig, M. and Leininger, W. (1988), 'Markov-perfect equilibrium in games of perfect information', Sonderforschungsbereich 303 'Information und die Koordination wirtschaftlicher Aktivitäten', Discussion Paper No. A-183.

Hillman, A.L. (1989), *The Political Economy of Protection*, Harwood: Chur.

Hoel, M. (1986a), 'Bargaining games with potential outside offers', Memorandum from Department of Economics, University of Oslo, No. 12.

Hoel, M. (1986b), 'Perfect equilibria in sequential bargaining games with nonlinear utility functions', *Scandinavian Journal of Economics*, **88**, 383–400.

Hoel, M. (1987), 'Bargaining game with a random sequence of who makes the offers', *Economics Letters*, **24**, 5–9.

Johnson, H. (1953), 'Optimum tariffs and retaliation', *Review of Economic Studies*, **21**, 142–53.

Kandori, M. (1992), 'Repeated games played by overlapping generations of players', *Review of Economic Studies*, **59**, 81–92.

Kennan, J. and Wilson, R. (1993), 'Bargaining with private information', *Journal of Economic Literature*, **31**, 45–104.

Klein, B. (1985), 'Self-enforcing contracts', *Journal of Institutional and Theoretical Economics*, **141**, 594–600.

Kreps, D., Milgrom, P., Roberts, J. and Wilson, R. (1982), 'Rational cooperation in the finitely repeated prisoners' dilemma', *Journal of Economic Theory*, **27**, 245–52.

Krishna, V. and Serrano, R. (1996), 'Multilateral bargaining', *Review of Economic Studies*, **63**, 61–80.

Kydland, F.E. and Prescott, E.C. (1977), 'Rules rather than discretion: the inconsistency of optimal plans', *Journal of Political Economy*, **85**, 473–91.

Laffont, J.-J. (ed.) (1992), *Advances in Economic Theory: Sixth World Congress*, Vol. 1, Cambridge: Cambridge University Press.

Ma, C.A. and Manove, M. (1993), 'Bargaining with deadlines and imperfect player control', *Econometrica*, **61**, 1313–39.

Martin, S. (1993), *Advanced Industrial Economics*, Cambridge: Blackwell.

Maskin, E. and Tirole, J. (1994), 'Markov perfect equilibrium', Harvard Institute of Economic Research, Discussion Paper No. 1698.

Mohr, E. (1988), 'On the incredibility of perfect threats in repeated games: note', *International Economic Review*, **29**, 551–5.

Mohr, E. (1991), *Economic Theory and Sovereign International Debt*, London: Academic Press.

Muthoo, A. (1995), 'Bargaining in a long-term relationship with endogenous termination', *Journal of Economic Theory*, **66**, 590–98.

Nash, J.F. (1950), 'The bargaining problem', *Econometrica*, **18**, 155–62.

Nash, J.F. (1953), 'Two-person cooperative games', *Econometrica*, **21**, 128–40.

Okada, A. (1991a), 'A noncooperative approach to the Nash bargaining problem', in Selten (1991), pp. 7–33.

Okada, A. (1991b), 'A two-person repeated bargaining game with long-term contracts', in Selten (1991), pp. 34–47.

Osborne, M.J. and Rubinstein, A. (1990), *Bargaining and Markets*, San Diego: Academic Press.

Pearce, D. (1992), 'Repeated games: cooperation and rationality', in Laffont (1992), pp. 132–174

Ray, D. (1994), 'Internally renegotiation-proof equilibrium sets: limit behavior with low discounting', *Games and Economic Behavior*, **6**, 162–77.

Rees, R. (1993), 'Tacit collusion', *Oxford Review of Economic Policy*, **9** (2), 27–40.

Roth, A.E. (ed.) (1985), *Game-theoretic Models of Bargaining*, Cambridge: Cambridge University Press.

Rubinstein, A. (1979), 'Equilibrium in supergames with the overtaking criterion', *Journal of Economic Theory*, **21**, 1–9.

Rubinstein, A. (1982), 'Perfect equilibrium in a bargaining model', *Econometrica*, **50**, 97–109.

Rubinstein, A. (1985a), 'A bargaining model with incomplete information about time preferences', *Econometrica*, **53**, 1151–72.

Rubinstein, A. (1985b), 'Choice of conjectures in a bargaining model with incomplete information', in Roth (1985), pp. 99–114.

Rubinstein, A. (1992), 'Comments on the interpretation of repeated games theory', in Laffont (1992), pp. 175–81.

Rubinstein, A. and Wolinsky, A. (1992), 'Renegotiation-proof implementation and time preferences', *American Economic Review*, **82**, 600–14.

Scherer, F.M. and Ross, D. (1990), *Industrial Market Structure and Economic Performance*, Third Edition, Boston: Houghton Mifflin.

Schultz, C. (1994), 'A note on strongly renegotiation-proof equilibria', *Games and Economic Behavior*, **6**, 469–73.

Selten, R. (1965), 'Spieltheoretische Behandlung eines Oligopolmodells mit Nachfrageträgheit' ('Game theoretic treatment of an oligopoly model with demand inertia'), *Zeitschrift für die Gesamte Staatswissenschaft*, **12**, 301–24, 667–89.

Selten, R. (1975), 'Reexamination of the perfectness concept for equilibrium points in extensive games', *International Journal of Game Theory*, **4**, 25–55.

Selten, R. (ed.) (1991), *Game Equilibrium Models III. Strategic Bargaining*, Berlin: Springer.

Shaked, A. and Sutton, J. (1984), 'Involuntary unemployment as a perfect equilibrium in a bargaining model', *Econometrica*, **52**, 1351–64.

Siebert, H. (1998), *Economics of the Environment*, Berlin: Springer.

Smith. L. (1992), 'Folk Theorems in overlapping generations games', *Games and Economic Behavior*, **4**, 426–49.

Stähler, F. (1996a), 'Reflections on multilateral environmental agreements', in Xepapadeas, A. (ed.), *Economic Policy for Natural Resources and the Environment*, Aldershot: Edward Elgar, pp. 174–96.

Stähler, F. (1996b), 'Bargaining in a long-term relationship and the Rubinstein solution', Kiel Working Paper No. 759, Kiel Institute of World Economics, Kiel.

Stähler, F. (1996c), 'Markov perfection and cooperation in repeated games', Kiel Working Paper No. 760, Kiel Institute of World Economics, Kiel.

Stähler, F. (1997), 'Strategic bargaining and the Nash solution', mimeo.

Stähler, F. and Wagner, F. (1998), 'Cooperation in a Resource Extraction Game', Kiel Working Paper No. 846, Kiel Institute of World Economics, Kiel.

Sutton, J. (1986), 'Non-cooperative bargaining theory: an introduction', *Review of Economic Studies*, **53**, 709–24.

Telser, L.G. (1980), 'A theory of self-enforcing agreements', *Journal of Business*, **53**, 27–44.

Thaler, R.H. (1988), 'The ultimate game', *Journal of Economic Perspectives*, **2**, 195-206.

Thomas, J. and Worrall, T. (1994), 'Foreign direct investment and the risk of expropriation', *Review of Economic Studies*, **61**, 81–108.

Tirole, J. (1988), *The Theory of Industrial Organization*, Cambridge, Mass.: MIT Press.

van Damme, E. (1989), 'Renegotiation-proof equilibria in repeated prisoners' dilemma', *Journal of Economic Theory*, **46**, 206–17.

Wen, Q. (1994), 'The "Folk Theorem" for repeated games with complete information', *Econometrica*, **62**, 949–54.

Index